*f*P

On the Origin of Tepees

The Evolution of Ideas (and Ourselves)

Jonnie Hughes

Free Press

New York   London   Toronto   Sydney   New Delhi

Free Press
A Division of Simon & Schuster, Inc.
1230 Avenue of the Americas
New York, NY 10020

First Free Press trade paperback edition June 2012

FREE PRESS and colophon are trademarks of Simon & Schuster, Inc.

For information about special discounts for bulk purchases,
please contact Simon & Schuster Special Sales
at 1-866-506-1949 or business@simonandschuster.com.

The Simon & Schuster Speakers Bureau can bring authors to your live event.
For more information, or to book an event,
contact the Simon & Schuster Speakers Bureau
at 1-866-248-3049 or visit our website at www.simonspeakers.com.

Designed by Ruth Lee-Mui

Manufactured in the United States of America

1   3   5   7   9   10   8   6   4   2

The Library of Congress has cataloged the hardcover edition as follows:

Hughes, Jonnie.
On the origin of tepees : the evolution of ideas (and ourselves) / Jonnie Hughes.
p. cm.
Includes bibliographical references.
1. Tipis—History. 2. Tipis—Social aspects.
3. Indians of North America—Dwellings. I. Title.
E98.D9H83 2011
303.4—dc22
2010051295

ISBN 978-1-4391-1023-2
ISBN 978-1-4391-1024-9 (pbk)
ISBN 978-1-4391-5805-0 (ebook)

To JJ,
with love

Contents

Part V: Mysteries Solved

Acknowledgments

I'd like to extend my thanks to Hilary Redmon of Free Press; Robert Kirby and Barbara Zitwer, for their encouragement and support; my sister Emma King and good friends Dave Forman and Andy Bedwell, for agreeing to read early drafts; Dr. Anna Prentiss, for valuable information on Great Plains archaeology; Montanans Mike and Kathy Kasic, Rick Mace, and John Jarvis, for their splendid hospitality; Gladys and Reginald Laubin, for *The Indian Tipi*; Mr. P. Veale, for putting me on this path; my wife, Julie, and children, Dael, Mia, Joel, and Max, for permitting me the space and time to follow it; my parents, Pam and David Hughes, for urging me ever onwards; and, of course, my brother Ads, for sharing the driving.

PART I

Only Human

1

Weirdoes

Misting Up

We all look at the world through goggles. Many of us are unaware of this fact (and those who are aware are loath to admit it), but we all perceive the world about us through tinted lenses—tinted with the ideas stored in our memories. Only by referring to the millions of ideas we have consciously and unconsciously logged in our brains can we continually make sense of the world we experience. No one has a goggles-free view of the world, because no living brain is ideas-free.

That said, we do have a choice of goggles. There's a large rack to browse. Hanging there, in no particular order, are libertarian goggles, Zen Buddhist goggles, Freudian goggles, environmentalist goggles,

Marxist goggles. We can take our pick. We can pick more than one pair if we wish.

Historically, I've tended to wear a pair that first appeared in 1858 and came free with every volume of Charles Darwin's book *On the Origin of Species*. By all accounts they caused quite a stir on release. Through them, the world looked godless and pointless. It took quite a while for the average pair of eyes to adjust to them, and even when they had, many didn't like the view. Nevertheless, Darwin's goggles persisted. They were worked upon and improved, and nowadays they are standard wear for many of those growing up in the Western world. I've worn them for as long as I can remember; I love the view. As I see it, a godless, pointless world can still be fascinating and wonderful—if not more so precisely *because* it is godless and pointless. But of course, I would say that; I'm wearing Darwin's goggles. However, even *I* can see that Darwin's goggles are not perfect. While they work wonders at bringing the rest of Life on Earth into sharp focus, when you look through them at our species, an odd thing happens: the lenses mist up. The goggles appear unable to reveal humankind with any great acuity. Through them, no matter how hard you squint, *Homo sapiens* looks, well . . . a bit fuzzy.

Let me explain . . .

The Top Five Weirdest Wonders in All Creation

In reverse order, these are:

5. THE HAMMERHEADED FRUIT BAT *(HYPSIGNATHUS MONSTROSUS)*
The Largest, Loudest, and Ugliest Bat In Africa.

Male hammerheaded fruit bat.

No bat is beautiful, but even in relative terms, the male hammer-headed fruit bat must be said to be butt ugly. Its face consists of a bulbous muzzle with pop-out green eyes on top and a drooping lip underneath. "Village idiot" is its default facial expression. Beneath its warty skin it is just as disfigured. If you were to open one up, you'd find lungs and an oversize larynx occupying the upper third of its body cavity, and two magnificent testicles filling up the lower. There is little space in there for fruit, because the *raison d'être* of a male hammerheaded fruit bat is not eating; it's honking and bonking. They are the randy brass section of the animal world. Their sole purpose in life is to play music that lures in groupies.

During mating season, the males gather at sunset in their favorite tree by the side of a river. Sound carries well over the water, but hammerheaded fruit bats need little amplification: these animals create an inimitable din. The sound they make has no parallel on Earth, but you can just about conjure it up in your imagination if you swap the whistles of a crowd of overzealous ravers for fictional claxon-bugle hybrid instruments and kill the trance music so that you hear only them. These inane toot blasts are repeated *ad nauseam* for hours on end, accompanied by frantic wing flapping, until the female hammerheaded fruit bats (almost pretty by comparison) eventually relent and visit their manic, honking conspecific to collect their sperm.

Although to human ears the horrendous clamor of one bat is indistinguishable from that of another, there is clearly some artistry involved in this noisemaking. A recent study found that 80 percent of baby hammerheaded fruit bats were sired by only a sprinkling of the males—clearly the best honkers. Bonkers!

4. THE OARFISH (*REGALECUS GLESNE*)
The Oddest of All the Oddities of the Sea.

Most people have never heard of this fish. Those who have don't necessarily believe it exists. It's the star of sea serpent legends around the world. In Norway, the Vikings called it the "king of herrings"; they imagined that it swam ahead of herring shoals of biblical proportions, guiding the masses through the depths and away from the Vikings' nets, its luminescent body shimmering in the dark like a beacon.

But the oarfish is a real creature, and one of Life's great mysteries. The world's largest bony fish, it can measure up to ten meters long and weigh as much as a quarter of a ton. Its tapering body has a coat of silvery scales as if it's been dipped in zinc. It has a king's crown of

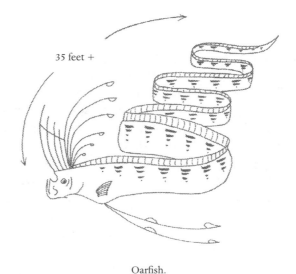

35 feet +

Oarfish.

bright crimson quills on its head, pectoral fins resembling six-foot-long red chopsticks—the "oars" in its name—and a dorsal fin that travels the length of its back like a picket fence.

No one is certain what the oarfish eats, where it lives, or how it breeds. While specimens have occasionally been washed up on beaches after storms or caught in nets, it wasn't until 2001 that anyone saw an oarfish alive in the sea. That year, while checking a buoy off the Bahamas, Bill Cooksey, a U.S. Navy diver, spotted one suspended just under the surface. It was hanging *vertically* in the water, its bizarre crest erect, its "oars" splayed out to each side. For just a moment, Cooksey thought he was looking at a giant silver crucifix glinting ahead of him in the gloom. When he realized that he was in the company of a genuine sea monster, he didn't feel any more at ease. As he approached the creature, it remained motionless, apart from its huge eye, which tracked his every movement. When he was within a few meters, it rotated on its axis, keeping its broad side to him, by sending subtle ripples along its dorsal fin. Cooksey was amazed at its control in the water; it moved like a machine. He actually got close enough to touch it before it decided to exit the scene. When it did so, it didn't dart off through the surface waters, but instead dropped downward like a stone into the darkness, its eye staring up at him as it went.

3. THE ELEPHANT *(FAMILY ELEPHANTIDAE)*
An *Extra*terrestrial with ESP.

Elephant.

I know they are familiar animals, but if you look at them again, as if for the first time, you'll see that elephants are quite alien. The first giveaway is that they are oversize for this planet. The most massive elephant ever recorded weighed in at twelve tons. That's three times as big as anything else that walks the Earth. Their impact on the environment is equally larger than Life. Elephants are the only animals (other than us) capable of changing rain forest into savannah, savannah into grassland, and grassland into semidesert. That's not to say that their "habitat engineering" is purely destructive. Elephant tree felling tends to increase species diversity by encouraging fresh flushes of plant growth; their path making opens thick bush to dozens of other herbivores; they are the single most important seed dispersers in Africa; and the tonnage of their dung is sufficient to feed more than three thousand kinds of invertebrate.

Second, their appearance is otherworldly: unshapely lumps of flesh covered in a bright gray, virtually hairless skin that is two sizes too big, and four legs that are little more than posts with toenails. African elephants have ears big enough to sleep under, and two incisors that never stop growing, so that by the time they are five years old, they need to sidestep trees to avoid being caught up by their own dentition.

Then there's the appendage on their faces. The prehensile snout is not unique to the elephant—the saiga antelope, the tapir, the aardvark, and the numbat of Australia all boast a grasping nose—but no other creature

has one as impressive. The elephant's trunk is a nine-foot-long elonga-
tion of a fused nose and upper lip. That's why the end of it looks like a
pouting nostril. It contains more than forty thousand individual muscles
so that it can orient itself in any direction. The trunk is so sensitive and
dexterous that it can locate and pick up a single blade of grass, yet it is
strong enough to rip branches off an adult acacia tree.

In terms of utility, the elephant's trunk is nature's multitool. In a
single day, an elephant trunk may dice a tree into portions small enough
for the elephant to chew; suck up fifty gallons of water in four-gallon
portions and spray them into the elephant's mouth; shake the fruit off a
tree; move logs out of the way; act as a snorkel in water to supply air to
the elephant swimming beneath; spray mud over the elephant to protect
its skin from the sun; greet other elephants with a trunk shake; rise high
in the air to sniff for predators, food, and other elephants; give close relatives and young a reassuring caress; wrestle with the trunks of other elephants to exert dominance; wipe mud from the elephant's eye; remove a
thorn from the elephant's foot; and trumpet to inform any other animal
within one mile that this elephant means business.

Behind the trunk is the sparkling elephant mind. Long-term studies
of wild herds suggest that elephants use tools, grieve for lost relatives,
possess an artistic aesthetic, show compassion, and are self-aware.
They also seem to be able to learn from one another's experiences.
Elephants in northern Mozambique can live in areas that have been
heavily seeded with landmines only because they have learned not to
step on soil that smells of TNT. This talent presumably arose when
elephants came to associate the odor with accidents that befell other
elephants. The fact that they can teach their young such new tricks
without incurring further calamity is a sign that elephants possess uncommon brainpower.

But perhaps the strangest feature of the elephant is its ability to communicate over vast distances. Katy Payne, a zoologist from Cornell
University, has studied elephants for twenty-five years. When she began
her work she discovered scientific reports going back decades that dared
to suggest that elephants had, to some degree, a capacity for telepathy.
They are able, the researchers insisted, to detect one another's presence
and communicate over a distance of several miles without making a
sound. Sure enough, Payne found that African elephants will spread out

over a huge area to move through the bush and yet they always appear to know exactly what the others are doing, or intending to do. It wasn't until she stood next to one in a zoo back home in the States that she realized how they did it.

"I felt a thumping in my chest that was exactly like the feeling I got when I was a child, on a Sunday, in my local church and sitting right next to the organ pipes as the organist played hymns. I was feeling the vibration of a sound wave so low, so huge, that my silly little human ears couldn't pick it up." Payne came back to the zoo the next day with a microphone and a tape recorder. When she played back her silent elephant song at high speed, there it was: a voice too deep for us to hear. Her work transformed our understanding of elephants. Today we know that elephants routinely use more than fifty different "infrasound" communications to organize their social lives. Because they have a longer wavelength, the vegetation hardly dampens the infrasounds as they advance, so they can travel immense distances. Forest elephants in the Congo can keep in touch with one another through tens of miles of thick jungle, singing through the undergrowth as whales sing through the oceans. An elephant can even use the ground like a telegraph wire, directing "seismic" rumbles from its body cavity down through its legs and, via *terra firma*, on to the recipient at the other end, who listens to the message with its foot.

2. THE NAKED MOLE RAT (*HETEROCEPHALUS GLABER*)
A Mammal with the Social Life of an Ant.

Naked mole rat.

Extreme environments call for extreme measures. One of the most extreme environments on Earth is found in the driest and most desolate parts of Ethiopia, Kenya, and Somalia, and these deserts have given rise to, for my money, the most remarkable but one living thing on Earth.

For much of the year the only signs of plant life here are pitiful collections of dry sticks emerging hopefully from dry soil. Yet deep below the surface, the plants exist as giant tubers full of calories and minerals. They propagate like strawberries, so where you find one, there will be others, and this clustering of such a good food source is just enough to support animal life. Evolution might have produced a large digging mammal tolerant of aridity, with an acute sense of smell, to harvest these scraps. Instead, it created a miraculous mob of burrowing animals so obscure that we're still rewriting the rule books as we discover more about them.

You can tell just by looking at them that they're (almost) the most unusual animals on Earth. Little more than autonomous penises with teeth at one end and comically oversize feet underneath, they may not be everyone's cup of tea, but don't scoff. Everything about the naked mole rat is bespoke. Their enormous buck teeth are mining picks, able to bite tunnels through even the most caked desert soils. Their podlike legs enable them to scamper about the colony at a hyperactive speed. Just like subway trains, they go backward as fast as they go forward, but unlike trains, when two of them meet in a tunnel, it doesn't cause delays. Naked mole rat etiquette is such that the subordinate mole rat will always crouch to allow the dominant mole rat to pass over the top. If there's some dispute as to who is what, they'll tussle for a while. Then the most bullish will walk over the other one.

But the naked mole rats' weirdness goes far beyond looks alone. They're no bigger than hamsters, yet they can live for up to twenty-five years. That's three times longer than any other mammal their size. They are the only mammals that can claim to be cold-blooded; they've lost the ability to control their body temperature. They are coprophagic—they eat their own feces—and if a mole rat is too fat to reach its own anus, it will beg droppings from that of a passerby. They can't feel pain in their skin; they lack the appropriate neurotransmitter. Even their nakedness is unusual: they are the only mammals smaller than a human to have lost their hair, and we've no idea why they did.

However, even if they displayed none of these characteristics, naked mole rats would still earn their number-two position on my list purely as a result of their unique social organization. In naked mole rat societies, all the females are sterile, apart from one: the queen. The queen alone gives birth to baby mole rats. She is a queen in the sense that a queen ant is a queen: she is everyone's mother.

But she's not the sort of mother to whom you'd send a Mother's Day card. Like the queen in *Alice's Adventures in Wonderland*, the mole rat queen is extraordinarily bad-tempered, walking (literally) over every other member of the colony. Since becoming a mole rat monarch is a career goal, not a birthright, her reign is vulnerable: at any moment one of the female courtiers below her may make a move to swipe her crown. The only way she can stay on top is to intimidate the rest of the colony into submission. That's why she's so bossy. She runs a reproductive dictatorship: her brashness halts the fruiting of her competitors' ovaries.

It's worth taking a moment to consider just how odd this situation is. This is a mammal that lives like an ant—in subterranean societies of sterile workers, soldiers, and a fertile queen—a social strategy termed eusociality, and certainly the most sophisticated design for animal organization found on Earth. Science is still reeling from the discovery that insects have come up with this ingenious tactic, but the discovery, in 1976, that a mammal had the same idea almost beggared belief.

This is because mammals are much bigger and more complicated "gene machines" than insects, and breeding within a close-knit society has extreme dangers for such creatures. Trapped in its remote subterranean tuber farm in the middle of the desert, any one naked mole rat colony is significantly isolated from the next. Hence, each mole rat queen is forced to adopt her close relatives as mates (her brothers, her sons, even her dad; nothing seems to faze her). As any biology teacher will tell you, incest is a very bad idea, for two reasons: First, it increases the incidence of bad genes, some of which are lethal, so that children of incest often don't make it to adulthood. Second, it reduces the genetic variety of the entire species, so that if its environment were to change, it would be unable to adapt and would soon be consigned to the fossil record. Yet molecular clock data suggest that naked mole rats have been scurrying about their tunnels shagging their relatives like Caligula himself for approximately thirty-eight million years; that's almost forty times the shelf life of the average rodent species. So how have they gotten away with it?

The patterns in mole rat DNA suggest that, at one point early in its evolution, the species did indeed enter a ruthless genetic bottleneck. This must have been the point at which incest became commonplace. Lethal genes must have circulated within the population like homemade viruses and done away with the majority of the species, but remarkably a small number of utterly inbred but genetically spotless mole rats seem to have survived. Equipped with a gene pool so shallow they hardly got their feet wet, these genocide escapees should have disappeared fairly abruptly the moment their environment fluttered. But the point is: they never did. The world of the naked mole rat has remained the same for almost forty million years. Quite by accident, this species has found itself in an "evolutionary vacuum." Huddled within its subterranean sweat holes, in the inalterable soils of African deserts, and oblivious to the messy world above, the naked mole rat has found a way of cheating evolution. It has stepped outside of Darwin's big project. There's only one other animal that is thought to have pulled off the same trick. . . .

1. US (*HOMO SAPIENS SAPIENS*)

Humans.

Life has produced some unimaginably weird living things, but our species, *homo sapiens sapiens*, the "wise wise man," really takes the biscuit. We are the only animal to talk. We are the only animal to walk upright in preference to any other form of locomotion. We are the only animal to cry as a sign of sadness. We have a larynx that is so far down our necks we can actually kill ourselves at dinnertime by choking on a piece of food. We are a social animal but have no optimal group size; we live in groups from 1 to 35.6 million.

Even among the primates, our closest relatives, we are oddballs. We are the only naked primate. We are by far the chubbiest primate; we have ten times more fat cells than any of our relatives. We are the oiliest, sweatiest primate. (No chimp gets acne.) We are the only primate to live successfully in the cooler regions of the world.

And then there is that enormous head. Our heads are so big that we are born prematurely in order that we may exit our mothers safely. As it is, maternal and infant death during childbirth is far higher in our species than in any other. If we make it to the outside world, as newborns, our heads are a quarter of our body length and a third of our mass. While our cousin the chimpanzee can happily lift its head within two weeks of birth, a baby human must wait twenty weeks before its neck is strong enough to hoist its oversize head up in the air.

Our heads are so big because our brains are so big, over three times as large as the brains of other animals our size. There are one hundred billion nerve cells in there, and more nerve cell connections than it will ever be possible to count. In IT terms, we each have 16,800 GHz of processing power, 1 million gigabytes of memory, and ports for up to 21 senses.* While the animals at the zoo or on your rug have pocket calculators in their skulls, we have supercomputers.

And we use them well. If Life dares to throw a problem our way, with a little time we can almost always come up with a solution. Our intelligence, above all, is the trait that most distinguishes us from the rest of the animal kingdom. I know people often say that dogs and dolphins are intelligent, or chimps—you can see it in their eyes, can't you?—but none of these animals comes within a mile of our *nous*, our common sense. Lassie, Flipper, even Tarzan's friend Cheeta—all are, in relative terms, stupid. They couldn't really do all those clever things in the films; they were trained to do them, by us. And they did them not for self-improvement or for the challenge or even for money, but to get another piece of ham, fish, or banana.

*Not just the five that Aristotle came up with.

Super-Natural?

Yet it is neither our life-threatening larynx nor our naked, spotty, sweaty skin, nor our dangerously large heads, nor our wonderful brains and sparkling intelligence that singles us out as the weirdest wonders in all creation. There is something even more peculiar about our species than all these things, and that is this: we are the only species on the planet that cannot be fully explained by Charles Darwin's otherwise faultless theory of evolution by natural selection.

Natural selection is powerful enough to explain how male hammer-headed fruit bats became flying trumpets, how oarfish came to look and move so oddly, how elephants developed their trunks and infra-sound, even how naked mole rat queens came to run their reproductive dictatorships. It does all of this without needing to call on some divine creator, because natural selection is a theory that can explain how nature makes marvels without intentional/conscious thought; it can explain how nature makes marvels mindlessly. Yet even natural selection, as we currently understand it, cannot explain how you came to be able to sit there and read this book; and that is because first, there's no call for you to be smart enough to read, and second, you shouldn't be wasting your time on this anyway when there are berries to pick and mates to bonk.

I'll elaborate . . .

1. THE MYSTERY OF OUR PAST: AN INEXPLICABLE EVOLUTION

For natural selection to work, a species has to have a hard life. Adversity is at the heart of all evolution because Life's innovations come only as a response to adversity. The environment—and by that we mean everything surrounding a living organism, from its neighbors to the weather—drives adaptation by selecting any trait that helps a living thing survive and reproduce, and extinguishing every trait that hinders either survival or reproduction. With nature selecting the fittest living things out of all those on offer, a species will adapt *until* it becomes "fit for purpose." No more, no less.

Humans evolved to survive and reproduce in the savannah of East Africa. Even the most optimistic evolutionary biologists concede that we are spectacularly over-equipped to do that.

BRAIN VOLUMES

Australopithecus	Homo erectus	Homo sapiens	Chimpanzee
350cc	800cc	1,350cc	450cc
3 million years ago	1.5 million years ago	Today	Today

Just look at the facts: it took just over three million years for us to evolve from a creature that looked a bit like a chimpanzee into the disfigured oddballs we are today: naked, sweaty, upright, chatty, brainy weirdoes. Our catalogue of traits went through the mill while our closest relatives hardly changed a bit. It was a bout of natural selection like nothing the world had ever seen—some strange driving force ballooning our brains from a moderately impressive 350cc to a distinctly overpowered 1,350cc. Would the African savannah ever need such a thing? We've got a supercomputer inside our heads and all we really needed to continue a perfectly respectable measure of surviving and reproducing—to enable our genes to make the short journey from one generation to the next— was the standard-issue pocket calculator our relatives had. The extent and the speed of our recent evolution make no sense.

2. THE MYSTERY OF OUR PRESENT: AN INEXPLICABLE LIFESTYLE

Speaking of genes, we humans treat ours with utter contempt, and that is not only unwise but impossible by the rules of natural selection. Other living things only ever *spoil* their genes: they spend all their time and energy keeping them happy, propagating them. It's one of the universal truths of modern Darwinism that whatever an organism does or has, it must ultimately benefit the genes in some respect; otherwise it wouldn't do or have it. Sometimes we have to look really hard to see the "gene's-eye view," but we've never failed to find it, except in the case of us humans.

We do, or have, plenty of things that don't appear to benefit our genes in any way. We grow beyond reproductive age, as though there were a point in doing so. We choose not to have children, as though we have license to. We die for our country. We become celibate. We waste our energy collecting useless things such as stamps, instead of useful things such as nuts and fruit. We waste our time migrating to the Caribbean, not to survive the northern winter but to turn our skins brown.

Admittedly, some of these things may benefit some genes in as-yet-unimaginable ways—indeed, this is the last remaining hope of a whole cohort of Darwin's children, the evolutionary psychologists—but it does appear that at least parts of our lifestyles are not just weird but "illegal" according to the rules of life.

This inability of natural selection to explain both our evolution and our lifestyles keeps scientists up at night. You have to remember that since its inception, science has been busying itself with the project of humbling humankind. We humans once thought we were God images living at the center of a divine universe. Since then, we've been consistently stripped down a peg or two. First Copernicus taught us that the Earth revolves around the Sun, not the other way around. Then Galileo added that the Sun was in the armpit of a minor galaxy hidden in a vast universe of galaxies. Then Darwin worked out that Life was not some divine plan, but instead a mindless automated process of Life form making and that, yes, our fears were justified: those new animals in the zoo, the apes, look like us for a reason.

However, just as this dethroning was approaching its climax, Darwin's goggles misted up. The failure of natural selection to explain how our minds were made and how our lives were lived halted the philosophical advance of science in its tracks and enabled a line to be drawn in the sand. To many if not most of us, it still appears that "human nature" is apart from the rest of the living world, a product not of Darwinian selection but of divine intervention. Our weirdness itself gives the idea of God something to hold on to.

So, is that how we leave it? While all the other living things around us, and even most parts of our own bodies, can be explained by Darwinism, built by natural selection, the odd bits that make us human must remain non-Darwinian, built by *supernatural selection*. True fans of Darwin's goggles would never allow such a thing. As the philosopher Daniel Dennett suggests, it would be like explaining the construction of a skyscraper by saying that the bottom two thirds were built by cranes and the last one third by "skyhooks": construction arms dangling down supernaturally from the clouds. If you believe in the supernatural, if you're religious, then that's fine; skyhooks work. But many don't believe in the supernatural. Many, like me, believe that *"natural"* is all there is. Surely *we* can't rest until we've found those missing cranes, the ones that built the last third of our species. But how do we do this?

By trying on a new pair of goggles. A pair that will enable us to look at, from a new angle, the only thing we know of that manufactures

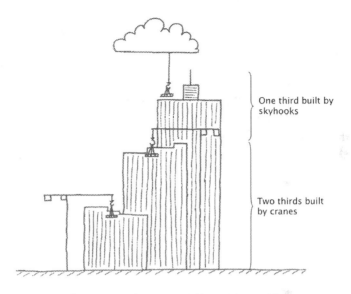

One third built by skyhooks

Two thirds built by cranes

A skyscraper partly constructed by supernatural forces.

cranes: natural selection. A pair of goggles that will present our evolution and our lifestyles in such a way that Darwinism will regain its foothold on our species. A pair that can spotlight the strange driving force that ballooned our brains, stripped our skin of its hair, stood us upright, and dropped our throats. A pair through which the human being does not appear over-equipped for its environment, but instead perfectly fit for purpose. A pair that can help us to see the adaptive value of "dual income, no kids."

I may have such a pair.

They're brand-new, still at the prototype phase—so new, in fact, they don't even have a name yet. I didn't make them. They're the work of a loose assembly of radical scientists and philosophers. They're very similar to Darwin's goggles in a lot of ways (and I'm sure he would have found them comfortable to wear), but they differ in that they are next to useless when looking at animals and plants. These goggles have been designed specifically for delivering a new perspective on *our* species and our species alone.

And, apparently, they're very good at it. The makers claim that with these goggles, the wearer will come to understand the human condition for the first time. At present, I'm not in a position to substantiate or refute this claim, because my eyes have only recently finished adjusting to their new lenses, but I have to say that it looks promising. These could be the goggles for me!

And they could be the goggles for you, too. That's why I'm writing this book. It is intended to enable you to make that judgment. I'm including a free pair of these new, nameless goggles with every volume. Let's put them on and explore the new world they bring into view together.

Step one: Let your eyes adjust . . .

2

The New World

The Great Indoors

Ads and I stand by, powerless, as Kenny G murders "The Girl from Ipanema." We're in an elevator with two men, one woman, and a toddler. The vista flashes before us, alternating glimpses of a vast interior space glittering with delights and touched-up I beams at close quarters. The lift is fast enough to make us all feel a little sick, and the woman next to us groans.

We're traveling within the bowels of the Mall of America (the MOA), the most visited enclosed shopping center in the world. There are more than 2.5 million square feet of retail space within these walls. Twelve thousand people work here. It has over 500 stores. There can't be much on Earth that you can't buy in this wondrous room. There is a store here that sells just beads, and another that sells just slippers. There's a dedicated remote-control helicopter store, a "Wallet World," a beanbag furniture store that sells "the security of a stuffed animal with the functionality of a pillow—It's a pillow and a pet. It's a Pillow Pet"—and a store that sells only the things people would need if they wanted to give up smoking. You couldn't count the number of different things for sale in this building. The diversity, the tiny differences between each different item—it's mind-blowing. And if you don't want to buy stuff, the MOA boasts a comedy club, an eighteen-hole miniature golf course, a staged wedding each night that you can buy tickets to, a shark tank that you can swim in, and a theme park complete with roller-coasters and log flumes—all of it under one roof.

What sort of people would need such an overzealous interior? The answer: Minnesotans. Ads and I are standing inside an inside, inside Minnesota. And Minnesota is a place in which being inside matters. This state suffers five months of snow and temperatures of between -60°F in winter and 114°F in summer. If you want to live a full life in Minnesota (and Minnesotans do), you have to build yourself a haven from the Great Outdoors. Your own little world. A Great Indoors.

The elevator pings, the doors open, and the other people shuffle out. Ads and I follow them into towering cloisters lined with purchasing opportunities, and discover that even at 10:00 AM, the mall is full of people of every description, shuffling, gazing, buying.

The Mall of America must be the preferred habitat of my species. There are probably tens of thousands of us in this room. If an ecologist ever stumbled upon a species as large as we are, in such a swarm, they'd have to conclude that they had discovered the species' hotspot, its idealized niche: a patch of the planet that just happens to satisfy all the needs of that particular living thing, fully and completely. Home Sweet Home. And they'd be right. The MOA does satisfy all of our needs. It's warm and dry. It has a superabundance of food and water. There are no natural predators here, and it's largely disease-free. It's the perfect place for a rather weird naked ape from tropical East Africa.

In genetic terms, we are all (roughly) East African. Our species, *Homo sapiens*, evolved there ten thousand human lifetimes ago, and although there have been some minor tweakings and additions since then, our gene pool, the totality of all our human genes, has remained (roughly) the same ever since. "Human evolution is over," says Professor Steven Jones, a leading geneticist at University College, London. After 5.8 million years of madcap, inexplicable evolution in which our brain inflated faster than an airbag in a head-on collision, our bout of natural selection seems to have come to an abrupt end with the advent of the "caveperson" in East Africa two hundred thousand years ago. Which means that, in genetic terms, Ads and I, and all the Minnesotans in this mall, *are* cavepeople. We may feel thoroughly modern, but our bodies are out-of-date by two hundred thousand years. If a genuine antique cavewoman entered The Gap right now and had the wherewithal to start flipping through the jeans next to me, I wouldn't bat an eyelid. She would look (roughly) the same as the other modern cavewomen in the store. I could have children with that antique cavewoman, our DNA would be that similar. So what's going on? Why do we cavepeople appear to be frozen in time?

Sandwich Selection

Ads and I gravitate toward the food court; we forgot to have breakfast this morning. I know, as Roseanne once said, "that's a special kind of stupid," but hey, we just flew in from London yesterday, so our clocks are off. I stand there, identical to all the other cavepeople, looking up at the options. No hunting or gathering for us anymore; we can do the money-for-goods swap that cavepeople do nowadays, and walk away with a hot, steaming quesadilla. Or shall I go for the bagel? It's healthier. (As if that's ever really mattered to this caveman: two hundred thousand years on, my Stone Age desires for protein, salt, and sugar still rule.)

Natural selection may no longer be happening to us, but it *is* happening to the sandwiches in the food court.* Every time one of us makes a choice, the sale is made, the stock runs low, and the manager makes a mental note to order more next time. The popular choices—those "selected for"—will be reordered (will *reproduce*) and grow in number over time. The dried-up specials that the manager thought were "worth a try"—those "selected against"—will not be reordered (will fail to

---

*Because human choice is involved, this is technically known as artificial selection, but bear with me.

reproduce) and will eventually become locally, if not globally, extinct. Over the weeks and months, people's selections dictate the sandwich "species" composition of the food court community.

But as well as selecting *between* species, food court natural selection works *within* species. Sift through the pile of supposedly identical tuna melt panini and you'll find some that are slightly older: The tuna looks a bit brown. There are some with a huge piece of raw onion sticking out; not good for the breath. One has clearly been dropped and hastily reassembled. These individuals are "less fit" than the younger, onion-meek, tidy ones. If tuna melts are sufficiently abundant—if there are more tuna melt panini than tuna melt choosers—the "selection pressure" will be sufficient to ensure that the less fit individuals remain in the refrigerator. A good manager will spot the problem and complain to the supplier: "Don't give me brown tuna, easy on the onion, and don't repack a sandwich unless you do it carefully." Over time, the population of tuna melt panini will adapt. Their "average fitness" will increase. In the human world, we call this process "quality control," but outdoors, in the wild, it is natural selection.

So how come our quality is no longer controlled? How is it that our two-hundred-thousand-year-old cavepeople bodies are never selected against, no matter how much metaphorical onion is hanging out of our metaphorical panini? How did we absent ourselves from biological evolution?

Remember the naked mole rat? It also managed to sidestep natural selection. How? It tucked itself away in a world of its own making, a perfect little "indoors" with a constant temperature and humidity year round. A safe place where cruel nature can't get at it, where there is no "survival of the fittest" to worry about. Well, we do that, too. We live indoors, only we go one step further: we take our *indoors* into the *outdoors*.

We don't huddle in the dark in a never-changing bunker. We venture out across the globe to deserts and mountain ranges and islands and Minnesota. We go to these extremely dissimilar places—places in which a naked tropical ape has no place being—and do something magical: instead of adapting to suit the local environment, we *adapt the local environment to suit us*. No matter where we are on Earth, we can fashion *Home Sweet Home*. The Mall of America is just the start of it. Outside here there are tens of thousands of farms and shops and public buildings and homes supplying everything we need: warmth, water, food, shelter, a lack of predators, the near-absence of disease. It's like we're forever living in an idealized East Africa, an East Africa that only ever means us well. And because we're always living in the best of all possible East Africas, our antique caveperson bodies never have any quality-control issues. Our "manager" never has cause to complain.

It all seems too good to be true, and if you're like me, a skyhook skeptic, the natural response is that it must be. I don't believe in magic. This "miraculous" power must be a conjuring trick. So what are we doing when we build Home Sweet Home in Minnesota? What's the trick?

Little Lars on the Prairie

This time, Ads and I use the stairs that border the enormous atrium in the middle of the mall, the home of the Nickelodeon Universe. I try to relax and enjoy the spectacle of an indoor Spongebob Squarepants rollercoaster, but I'm cross because I've been in the States for less than twenty-four hours and I've already ruined one pair of pants. I knew I shouldn't have gone for that tuna melt panini.

Barnes and Noble sucks us off the lower cloister. Bookstores have a tractor beam effect on me. Inside, near the Local History section, a couple starts talking to their children in a strange language, and this takes me totally by surprise. Ads and I had the distinct impression that we would be seeing only Americans in Minnesota. Yet here are people, deep in the mall, speaking an exotic tongue. The language is guttural but not unpleasant. A melody loops around the words, and as the mother speaks, I join her children and patiently watch her lips move, entranced. When I realize that they are all now staring at me, I glance back at the shelves, notice the numerous books on Scandinavian immigrants, and the whole thing comes together. They must be speaking Swedish or Norwegian. They *are* Americans—*Scandinavian* Americans.

Over a quarter of the people in this mall will have Scandinavian ancestry. Norwegians, Swedes, Icelanders, Finns, and Danes began coming to Minnesota as early as 1825, three decades before it was officially a state. They left Europe because it had become a place of political unrest, bad harvests, and religious intolerance. Upon landfall in the busy East, grabbing land in the Far West seemed the only option. Naturally enough, these people from the North of Europe gravitated toward the North of the United States. At that time, Minnesota was the most *northern* and *western* you could get: a cold, forested wilderness sparsely populated by French fur trappers, woodland Indians, and the results of their occasional union, the *Métis*. Fur was big business back then, and the young United States resented my nation, Britain, because it had a tight control on the trade: a grasping hand that came down from the North, stole the pelts from the backs of American animals, and retreated again to get rich. As an act of intent, the U.S. government built a castle, Fort Snelling, where the Minnesota River meets the Mississippi, only eight miles from where I am now. The immigrant Scandinavians hurriedly gathered around this northwest outpost.

Minnesota was much less habitable in the early nineteenth century, before there were nice, warm interiors to hide in. Starting from scratch in a place like this was tough. However, there were some advantages to the location: the falls just above the fort were a superb source of power for timber mills, and the wide river just below the fort marked the highest navigable point on the Mississippi. Within a few years, a milling industry had been set up on the falls to process the trees as they were taken down, and a river port had budded on the banks downstream. By the middle of the century, the mills had given birth to Minneapolis, and the port had become St. Paul, together the Twin Cities of Minnesota.

As the woodland fell, the Scandinavians set up farms on the prairies. As the mills turned from timber to flour, the settlers fed them with grain. To their great fortune, they had happened upon a young, rich soil that had been laid down only as the glaciers retreated, ten thousand years before. It was as productive as any in North America. The good soil began to burst with wheat, and Minneapolis became the world's largest flour milling plant. Its prosperity drew successive Scandinavians from their homelands, especially Norway. In the century after 1825, one third of Norway emigrated to the United States. With the exception of Ireland, no single country ever emptied a larger percentage of its population into the United States. Minnesota filled up with Vikings, people perfectly equipped to conquer a wild land that spends up to five months of the year below freezing.

But in what way were they perfectly equipped? True, they had very pale faces, which would have helped them make enough vitamin D under the dull Minnesota sky,* but other than that, their genes hadn't given them any special tools for their new life in America. Instead, it was their fourteen-thousand-year history in a land of snow and lakes and cold plains and trees that made the Scandinavians fit for Minnesota. Nineteenth-century Scandinavians were born lumberjacks, because they had always been surrounded by forests. They were expert carpenters. They were world leaders in vernacular architecture—the craft of making beautiful, functional buildings out of local materials—because their small, dispersed population in Scandinavia had ensured that the region never fell into the riot of feudalism and the desire for the stone fortresses and flamboyant palaces so characteristic of the rest of medieval Europe. For six thousand years they had been farming lands with a short growing season. They already knew how to raise crops quickly in a poor summer. They unthinkingly built big barns to house their livestock in the winter. They automatically collected the winter manure and added it to their soils. During the cold, dark months, they didn't waste their time; they had a niggling feeling that they ought to be making clothes and tools. Their Viking days had given them long traditions of furriery, leather and metalwork. To top it all, these new Americans were psychologically and socially buttressed by a deep Lutheran tradition that denied extravagancies and frivolities and promoted hard, simple work; good, solid craft; and a strong, supportive community. They were preconditioned for life in Minnesota. Succeeding in this climate was already routine for the Scandinavians: they'd spent millennia learning how to survive in a place like this.

So it didn't matter that, underneath, they were just naked tropical apes. Within a few decades of arriving in Minnesota, the Scandinavians were safe and warm in this wild, cold place. Just like the naked mole rats, they had succeeded in constructing Home Sweet Home, a perfect little world of their own, an "indoors" where nature couldn't get at them. But unlike that of the naked mole rats, this shelter was not fashioned from the hard desert soil. It was built out of something, on the face of it, far more fragile: their memories of the things their parents and grandparents had always done.

---

*Vitamin D is made by the skin with solar power. The light skin of Northern Europeans and northern Asians evolved to compensate for the dullness of their homelands: it let more sunlight through than the original brown skin did.

And there's the magic, right there, hidden in the brains of us humans: a capacity to memorize another human's solution to a problem. This is, to put it mildly, an uncommon talent, and it sets us far apart from all other living things. To demonstrate exactly how, here's a quick survey of Life's approaches to problem solving. (Don't worry. It won't take long. In the 3.5 billion years that Life has been evolving, it has come up with only four of these approaches.) My framework for this venture is the classic playground question "Why did the chicken cross the road?" Only, here, the question is not "why" but "how," and we'll have to start with something much dumber than a chicken.

The Road to the Ultimate Problem Solver

Scenario: On one side of a road are four creatures: a jellyfish, a sea slug, a chicken, and a human. Each of them will try to solve the problem of crossing the road. I don't know *why* they do it, but I do know *how*:

HOW DID THE JELLYFISH CROSS THE ROAD?
(THE "DARWINIAN" CREATURES)

A jellyfish, one of the world's simplest creatures, is a good example of the first type of living thing that arrived on the Earth, the *Darwinian creature.*˙ These creatures have only one solution to any one problem: the solution that is hardwired in their coordination systems. Like most animals, the jellyfish is coordinated by nerve cells, or "neurons," laid out in networks, just like a railway system, so that each piece of track almost touches the next piece of track. When a stimulus is perceived, a nerve impulse, an electrical spark, is triggered at an origin "station." It travels down the track at an impressive speed and, upon reaching its terminus, initiates a response that can be only one of two things: the contraction of a muscle (to move the creature) or the release of a hormone (to alter the "settings" of the creature). When you boil it down, all animals, including us, are restricted to this modest choice of two, which means that all animal coordination is just a question of when to twitch muscles and when to squirt hormones. The key to getting more complex creatures is to give them not more response options but more complex railway systems.

---

*Plants, fungi, and all those microscopic living things that we can't satisfactorily classify also fall into this first category of creature.

The railway system of Darwinian creatures is basic: a series of simple routes that are laid down before they are born (which is what "hard-wired" means). There is no track building allowed during their lifetimes: they are stuck with what nature gave them, at the mercy of their innate "reflexes." If their network is poorly built, then they are destined to give the wrong response and possibly end up dead. The savior of Darwinian creatures is that each species will have many members, and each member will have a slightly different railway system and, therefore, a slightly different response to any one situation. This variation *within the species* is the key to the survival of the Darwinian creature. Among them, because they vary, there should be at least a few that can respond well to any one situation. And, by the rules of natural selection, it will be these fit individuals who survive to create the next generation. Perfect should the situation arise again. Not so great if a completely novel situation should come along. Such as crossing a road.

There stands our jellyfish on one side of the road. (Okay, we'll have to partially submerge the road in seawater so it can move.) It's joined by a dozen others of the same species. They look just like our jellyfish, but each is wired up slightly differently. When the gun goes off, one of them shoots across the road, numb to the inevitable tremors of the oncoming cars. It doesn't last long. Others move off more slowly, and one or two get squished before we begin to see that some are hardwired to freeze when they feel the vibrations. If they are not in the line of the tires, then that's a good reflex. If they are in the line of the tires, then it's not. As it happens, the traffic on my road is quite heavy, so none of these eager jellyfish makes it.

(Our jellyfish still treads water on the edge. What is it waiting for?)

More jellyfish set off. One was born with a peculiar zigzag swimming pattern, which might have worked if it was swimming *toward* the vehicles, but moving perpendicular to them, it soon gets splattered. One continually head-butts the road surface—clearly it's preprogrammed to take to deeper waters, but none is available here, so . . . *squelch*. Now only two are left, our jellyfish and another, still wafting in a stable position at the starting line. For hours they wait by the side of the road, until nightfall. Jellyfish can detect the presence or absence of sunlight. These two individuals are hardwired to move only when the sun is switched off. It's just a fluke, but it will help them in this situation, because the traffic dies down after dark. Off they go. Lady Luck still has to play her part, but, sheltered by the increasing darkness, they both finally make it to the other side. Lone survivors of the road problem, these two individuals will be solely responsible for making the next generation of jellyfish. The population they found on the other side of the road will, more than likely, tread water all day and swim only at night, an "adaptation" that may be entirely pointless in their new environment. But that's Life.

The important thing is that no jellyfish ever gets to *make a decision* about what it will do next. It only ever acts in the one way that it can act: by playing out the behaviors that its equally oblivious parents once did in the same situation. Since they survived long enough to become parents, it's a good bet their hardwired behaviors will work again. This is why they are called Darwinian creatures, because natural selection is directly responsible for crafting their coordination systems. They will be forever dumb, because there is no cause to be anything else.

Now, don't be smug; there is more than a touch of the jellyfish in you. Ultimately, we stem from Darwinian creatures, and the proof can be found everywhere in our railway networks. Our eyes blink, our saliva gushes, our bladders fill and empty with or without our permission. We don't get to decide to have these responses. They are hardwired, innate reflexes straight out of a jellyfish, and perfect for the things that you don't need to think about.

So the answer to the question "How did the jellyfish cross the road?" is *by being lucky enough to have good reflexes.*

HOW DID THE SEA SLUG CROSS THE ROAD?
(THE "SKINNERIAN" CREATURES)

Burrhus Frederic Skinner, the American psychologist, discovered this kind of creature. Skinner experimented on pigeons, but we now know that even simple things such as sea slugs can work in this way. In fact, most living things fall into this category of creature. They differ from their ancestral Darwinian creatures in one important respect: they can learn.

When faced with a new problem, a sea slug will have not one but a number of different behaviors at its disposal. They pick one at random; no "thought" is involved. But if their behavior leads to a reward (in this case, not getting killed), the sea slug will be *more likely* to do that behavior should the same situation arise again. If they suffer as a result of carrying out a behavior (perhaps, in this case, experiencing a near miss), then they will be *less likely* to carry out the same behavior next time. In other words, sea slugs learn from their triumphs and disasters. They don't think about it beforehand, but they have a memory that can remember what happened afterward. To extend the railway network example, for every origin station, a sea slug offers a series of termini. Faced with a stimulus, it will send a spark down one of its routes *at random*. If the result is favorable, it will promote that route from a branch line to a main line. If not, it will close the line completely and try another route (assuming it survives the first attempt). Hence the railway network of a Skinnerian creature is able to respond to experiences: it has a certain degree of "plasticity," and plasticity is at the heart of all learning.

This simple type of learning is casually referred to as "trial and error." It's the foundation of most forms of animal training, and we humans routinely use it, too, when faced with new situations.

How did the sea slug cross the road? *Also by being lucky . . . but next time it won't need luck.*

HOW DID THE CHICKEN CROSS THE ROAD?
(THE "POPPERIAN" CREATURES)

Being a Skinnerian creature—being able to learn from your mistakes and triumphs—is extremely useful, but there's always the danger that you might choose the wrong opening gambit and get yourself killed. Far better would be a system in which you somehow carried out the trial-and-error process *in your head* beforehand. In other words, think about the problem and solve it before you do any behaving at all. These are the Popperian creatures, since it was the great philosopher of science, Sir Karl Popper, who said that foresight permits "our hypotheses to die in our stead." Popperian creatures are a subset of Skinnerian creatures, which are, in turn, a subset of Darwinian creatures. If you are a Popperian, you are both Skinnerian- and Darwinian-enabled. One example is a chicken.

In railway terms, a chicken faced with the daunting and novel prospect of crossing a road first "imagines" what would happen if it sent sparks down each of the available tracks. To do this, it must have *a model of the universe* in its head—a model that is realistic enough to generate useful predictions. This model will be a complete hodgepodge, constructed from the chicken's memories of all sorts of different learning experiences. It's probably not very good—the result of some confused and imprecise track laying—but if it enables the chicken to make a better-than-chance selection of what to do next, then it serves its purpose, and skill, not luck, decides the outcome.

Popperian creatures vary tremendously in quality, because putting together a good model of the universe is a difficult task. You have to re-member experiences accurately, you have to remember the correct "caus-ative" components of each experience, and you have to slot these useful experiences with great skill into your model of the universe. We humans are supremely talented at doing all of these things, orders of magnitude better than a chicken. We have big brains bulging with memory cells. We are exceptionally proficient at deducing the causes of effects that we experience. And we construct exquisite, intricate models of the universe inside our heads as no other creature can. But it's not completely effort-less. Every time you stop to ponder something, you can feel your brain working on it, can't you? Finding a good route on a map, considering your next move in chess, trying to pack the car—all involve referring to the model of the universe inside your head so that you can make a bet-ter-than-chance first move. It takes effort and care. And it's not foolproof; people get run over every day.

How did the chicken cross the road? *By coming up with a plan.*

HOW DID THE HUMAN CROSS THE ROAD?
(THE "DENNETTIAN" CREATURES)

Fortunately, on many occasions, we humans don't even have to come up with a plan, because we are Dennettian creatures, named after Daniel Dennett, the American philosopher to whom I owe this entire section; it's based, up to this point, on his Tower of Generate and Test, a model that describes the ways different brains react to the problems they en-counter. Dennettian creatures can do something far more impressive still than Popperians. We can solve the problem of how to cross the road safely without ever having experienced that situation in our lives before, and we can do it without taking any time to think about it. "How?" you ask. Simple: someone tells us how. Dennettian creatures are able to "bor-row" the lived experiences of other members of their species. They can either watch or listen to or read about the experiences of fellow Dennet-tian creatures and then make use of *their* solutions to Life's problems. In short: Dennettian creatures cheat; they swap thoughts!

While all those jellyfish are chancing that their immediate ancestors have survived similar situations, and the sea slug is crossing its proverbial fingers and plumbing for one of the few behaviors it can muster, and the chicken is standing there thinking about what to do next, we, the *crème de la crème* of Dennettians, can just yell across the road to someone who has already solved the problem and *get their thoughts* on the best way to do it. "I should go to the top of the hill if I were you. The cars slow down as they climb; you'll be able to see them for miles, and there are far fewer jellyfish."

Future-proof

Imagine how a group of such creatures will conquer a new environment. Moving into the new space, they will come across two types of problems: problems they have encountered before, and problems they have never encountered before. If they come across a novel problem, because they are Dennettian creatures, a subset of Popperian creatures (and hence a subset of Skinnerian creatures, and hence a subset of Darwinian creatures), they have at their disposal the three traditional methods of solving novel problems. Perhaps an instinctive reaction (*à la jellyfish*) will reap dividends; or maybe a spot of random trial and error (*à la sea slug*) will get somewhere; or possibly a quiet moment to think the problem through (*à la chicken*) will save the day. Obviously, we like to imagine that, in the case of us humans, a sophisticated Popperian approach is the default option. But let's not deny it: we frequently drop to the methods of our sea slug and jellyfish cousins. Just stand by and watch someone attempt to build Ikea furniture.

Regardless of the method by which each novel problem is solved, once it *is* solved, the solution instantly belongs not just to that individual Dennettian creature, but to the whole group (as long as the individual is prepared to share their thoughts). It is stored along with all the other learned solutions in what some would call a "collective memory," a repository of answers to Life's problems that all members of the group have access to.

Now imagine that repository. Over time, triumphant individuals will add *new* solutions; disappointed individuals will cast away *out-of-date* solutions that no longer work; and every individual will participate in the project of *improving* solutions. Each time the memory of a solution is put to work, a flash of on-the-job inspiration may make it work even better, or a hardwired quirk or a random accident may prove to yield a lucky addendum. All this change and revision in the body of solutions results in something profoundly significant: *the repository itself adapts to fit the environment.* While the Dennettian creatures feast on the spoils of their problem-solving genius, their collective memory takes on the job of interacting with the Great Outdoors. And because, as a body of thoughts, that collective memory becomes greater than any one of its hosts can handle, it starts to take on a superphysical existence; it begins to operate beyond the individual creatures that rely upon it. In short: the collective memory takes on a Life of its own. A Life that evolves.

This is our species' conjuring trick. We, as supremely able Popperian creatures, with outstanding models of the world in our heads, have managed to solve all the immediate problems that the outside world has thrown at us, problems that would have prevented us from surviving and reproducing. And because we are also the *crème de la crème* of Dennettian creatures, we have been able to share these solutions freely among our kind, so that the entire species is released from the hassles of natural selection. The result is that the human genome no longer has to deal directly with Mother Nature. These days, it works through an agent, an agent that negotiates on its behalf. I've referred to this agent as a collective memory, a repository of solutions, but there's a more common term for it. We call it culture.

It was the genius of the Scandinavians' culture that kept them warm and safe and well fed in wild Minnesota. It was their culture that evolved so that their naked tropical ape bodies wouldn't have to.

It may have taken 3.5 billion years of research and development, but eventually Mother Nature did it: she came up with the ultimate problem solver, the human being, a creature that solves problems not in the conventional sense, by adapting its biology, but in an unconventional sense, by adapting its culture. In so doing, Mother Nature achieved a design first, the goal of any technical engineer: she created a future-proof product, a product with "hardware" so sophisticated that it required no further work. All it would ever need to take on the future was upgraded "software."

A World of Our Own

How weird are we! By ensuring that the stuff in our heads is bang up-to-date, we've been able to keep hold of our cavepeople bodies, bodies that are two hundred thousand years out-of-date! These days the only parts of us that are engaged in a kind of evolution are the changing thoughts that fill our minds. We have (accidentally) swapped one type of evolution for another: biological for cultural. It's evolution, Jim, but not as we know it.

How do we put this into context? Where can we weirdoes, with our strange, perhaps unique evolution, place ourselves in the universe? Doesn't everything have its place?

Undoubtedly we belong to the part of the universe known as Earth, but we need more than that, because Earth is in fact a planet of many worlds. It has a world of gases, the "atmosphere"—the sum of all its gassy molecules, whether floating high above Everest, trapped in the lungs of a plunging sperm whale, or harboring in your rectum. It has a world of rock, the "lithosphere"—the global whole of all its natural solids, whether molten in a volcano, crushed way beneath Greenland, or sitting as pebbles on a beach. It has a world of water, the "hydrosphere"—the totality of all water molecules, whether frozen in the Ross Ice Shelf, pouring down from the skies over Oregon, or rising in the steam of your latte. These worlds are not entirely separate from one another, they're not "closed systems," but they do largely keep themselves to themselves, and each has its own particular suite of characteristics. We don't belong to any of these.

The world that we belong to wasn't even noticed until relatively recently. (They may be all around us, but the Earth's "worlds" are easily missed; you have to stand well back before you can get a good look at something as big as a world. Sometimes you have to squint.) It was only when Darwin published *On the Origin of Species* that our world, the biosphere, popped into view.

The biosphere is the world of Life: the sum of all genes, the global whole of Earth's ecosystems, the totality of all living things. It's a big, buzzing place. Mother Nature runs it, and she uses natural selection to quality-control the things that live in it, even the really weird ones. The hammerheaded fruit bats, the oarfish, and the elephants are all subject to its rules. Even the naked mole rats, if they ever dared to come and join the *real* world, would have to play the game. In fact, every last living thing appears to tow the line—except us. As I've said before, we're a bit of a conundrum.

There's no doubt that we are beings of the biosphere, too. Our physical selves, our bodies, are "biospherical." That's why we have aches and pains, teenage spots, bursting bladders, hunger, sex drives, childbirth, wrinkles, gases in our rectum. All of these are evidence of our membership in the biosphere. But aren't we more than just a collection of bodily functions? Aren't we more than our biology? We're certainly beyond at least part of the world of Life, beyond its natural selection. So how do we fit into the biosphere?

Well, that's the crux of the matter: we don't *exactly*. Uniquely among living things, while we are certainly a part of the biosphere, the biosphere is only a part of us. We alone appear to span *two* worlds. There's the old world, the biosphere, which sometimes we'd rather not admit to be in, and then there's another, *new world*, a world of our own.

This new world also lay hidden, right there in front of our noses, until 1926, when a little-known Russian geochemist from the early Communist era named Vladimir Vernadsky caught sight of it. And what he saw was this: "the world of human thought," the sum of all our memories, the global whole of our cultures, the totality of all known things. It was a superphysical world, manifest in all our artifacts—from pots to clothes to novels to cathedrals—but existing in truth beyond that, only as the continuing tiny firework displays in the collective of human minds. He had caught sight of something that was incredibly hard to see because this place was, well, otherworldly, literally *all in the mind*, or, more accurately, in all of our minds. This is the place in which our "human" selves meet, up above the feasting, farting, frigging biosphere. It was a big deal, so he came up with a name for it. He called it the "noosphere."

It's not a great name—it's from the forgotten Greek word for mind, *nous*—and that, along with the fact that he was writing in Russian at a time when Russia *was* a closed system, may account for its obscurity. For our discussion, however, the noosphere is a vital concept. In one word, Vernadsky was trying to conceptualize the grand repository of our entire species—all those solutions to all those problems—the world that opened up the moment we became Dennettian creatures and began swapping thoughts and building cultures. It must have started out very small, housed only in the uniquely bright minds of our ultimate ancestors in Africa ten thousand or more human lifetimes ago. But because of the way culture works, it would have grown exponentially and soon become too big for any one mind to handle. It would have taken on a Life of its own. Now it is a bona fide "-sphere," a whole new world. It covers the planet, because we cover the planet, because the cultures it enabled us to develop allowed us to cover the planet.

Into the New World

Deep in the out-of-date, old culture of the Scandinavian Americans— among the recipes for lutefisk and rutabaga, the sowing patterns for the bunad, and the best way to craft sled blades from antlers—are faint half-recollections of old stories, stories that had a precarious journey through the centuries and across continents: parceled by lips, delivered to ears in old tongues, captured on sheepskin parchment in a tangle of poetry and prose, then left to become decrepit on the oak shelves of royal libraries. These stories tell of a great hall, Valhalla, "the hall of the slain," a heaven run by a god for the martyrs of Viking battles so that, in death, they would receive everything they desired in life. Within its walls, Valhalla satisfied the needs of all its heroes, fully and completely. The feasting, the drinking, the entertainment, the indulgence—it never stopped. It was a 24/7 endless divine interior space. A room crammed with the happy immortal souls of once-mortal bodies.

It's tempting to conclude that, in the Mall of America, these Scandinavians drinking, entertaining themselves, indulging all around me now have their Valhalla, their heavenly hall—perhaps "Mal-halla," a 2.5-million-square-foot testament to the genius of their culture, parked on the spot where, less than two hundred years ago, they huddled as refugees on the edge of a new nation. But as Vernadsky discovered, the true Valhalla, the "great indoors," is larger than the MOA, than Minnesota, and than the United States. It's the heavenly hall of all human thought, the Big Repository to which we all have a key. While our mortal bodies are destined to plod on in the neighboring biosphere, released from the pressures of evolution yet still burdened with ill health, yearnings, and annoyances, our minds have made it to the noosphere, a place in which they can achieve immortality (if they ever manage to come up with something worth remembering).

My suspicion is that the weirdness of our species can be explained if we can understand this new world better, and here's my reasoning: If you knew nothing of water, how could you explain a goldfish? What would you deduce its fins, gills, and streamlined body were all for? If you were ignorant of the world in which a goldfish lived, everything about it would seem odd. So it could be with us. Perhaps we are having trouble explaining ourselves because we still can't see the bowl that surrounds us and the stuff we're swimming in. The bowl is the noosphere, and the stuff we're swimming in is culture.

So how do we get to discover the stuff we're swimming in? Why, by using goggles, of course.

There you go! You've caught up with me. Your eyes have adjusted, you've pulled focus, and the new world has come into view. But these new goggles should be able to do more than that; this is just the beginning. The claim is that we can use them to explore this new world and, in doing so, solve our species' unique mysteries. That's step two.

I can't wait to get started, and to that end, I'm standing at the gates of Mal-halla tap-tapping my heels, waiting for my brother, Ads, who's nipped to the bathroom to empty his biological self of excess urea. Looking down, I find that I've (accidentally) bought an *Encyclopaedia of Native American Tribes*. Adam has gone for a book on shamanism. He emerges. We take one last look at this heavenly hall, then head into the great outdoors.

PART II

What's the Idea?

A Revelation in Four States

3

Evolution, Minnesota

Is the Force with Us, Always?

The essential difference between these new goggles and all the other pairs on offer is that through these, culture is not simply an inanimate by-product of human activity, but is alive in its own right. We're not talking in metaphors here. For these goggles to work, we must entertain the suggestion that culture has a literal Life of its own—with a capital L—that the noosphere is *alive*, just as the biosphere is alive. So what is the meaning of Life?

Unfortunately, at present, we don't know. It is to the enormous frustration of every biologist that they cannot quite explain what biology, the study (*-ology*) of "Life" (*bios*), actually studies. When I was at high school, my biology teacher, Mr. P. Veale, realizing that he had to say something about the word, resorted to proposing that "Life" could be distinguished from "non-Life" in that those things that have "Life," *living things*, display seven different characteristics. They:

1. eat food of some sort (nutrition);
2. use energy (respiration);
3. respond to the external environment (sensitivity);
4. move in some way* (movement);
5. make waste products (excretion);
6. add to themselves (growth); and
7. can make other *living things* (reproduction).

I was among clever school kids, so this sevenfold definition, written up on the board for all to scrutinize, lasted about five minutes before someone said, "What about fire? Is that living, then?" Another added, "And crystals; they do all these things." They were right, and university science teachers know this, so to avoid embarrassment, they add subclauses. To point 2 they add "which they store in the molecule adenosine triphosphate." To point 6 they add "from the inside." This nips the smart alecks' argument in the bud but all seems a bit pathetic. Do we really have to be so specific in order to define "Life"? Does it really all come down to the presence or absence of adenosine triphosphate? Isn't it simpler than that? Isn't Life just, you know, Life?

If Mr. P. Veale had asked us school kids to explain what Life is at the start of the lesson, we wouldn't have come up with a list; we would have attempted to put into words a feeling that appears to be intuitive for most human beings: that "Life" is a kind of energy that exists in something when it is alive but that leaves when it isn't. In our clumsy sentences, we would have been pronouncing the central tenet of vitalism, a school of thought that arose spontaneously a long time ago in places far, far apart. The Chinese vitalists called this energy Qi. The Hindu vitalists called it prana. In *Star Wars* they called it the Force. It's supposed to exist behind and within every living thing. It's supposed to be the special ingredient, the pixie dust, of Life.

While we intuitively feel that the Force, or whatever you'd like to call

*Even plants that look static on the outside have busy little cells.

it, is there somewhere, it has never been found. To those who spend their days looking at living things at a magnification of one million times, vitalism is simply naïve. Scientists suggest instead that if Life is a thing at all, what we conceive of as Life must be an "epiphenomenon," a side effect of a more mundane activity that goes on in the universe. Distinguished scientists from a number of fields have drawn up contenders for this Life-affiliated activity. Unhappily, their suggestions all sound dry and, well, rather unsexy. Obi-Wan wouldn't approve. "Life results when autonomous agents self-organize in order to self-produce." "Life is what happens when a system works to reduce its internal entropy." "Life occurs when a multi-agent system exchanges materials and energy with its environment in order to complete one thermodynamic work cycle."

Hmm. Technically, these are true. But the point is that they are *all* true, so which one is the defining characteristic? Maybe there isn't just one. Perhaps you need all three and who knows how many others to trigger the Life epiphenomenon that we think we see. It doesn't feel like an answer. So here's another tack . . .

In a lab in San Diego, Professor Gerald Joyce is trying to get molecules of RNA, the baby brother of DNA, a little string of chemicals found in all our cells, to trigger a particular reaction. To some degree the RNA can already trigger the reaction, but they are slow and sloppy, and Joyce wants them to be quicker and more efficient. To improve their performance he can take one of two approaches. Either he can spend a lot of time analyzing exactly how the RNA triggers the reaction, and then go about the hard work of tweaking each little implicated atom in order to make it work better; or he can forget all that and let the RNA *evolve* the solution themselves.

"The recipe for Darwinian evolution of molecules is simple," Joyce says. "One, start with a large population of molecules of varying composition; two, select those molecules, however rare, that have the desired properties; three, produce many copies of the selected molecules, introducing occasional random changes in their composition; four, repeat as desired." With Joyce playing the hand of God, selecting only the best performers of each RNA generation to become the parents of the next, the RNA gets better and better at triggering the desired reaction over time. And because RNA is such a simple thing, it doesn't take long. Joyce can turn over a hundred RNA generations in a day. By the end of his working week he's five hundred generations down the line and the proud owner of RNA molecules that are superbly accomplished at their chore. The crucial point is that Joyce *doesn't even have to know how they are doing it.* In stark contrast to the approach he would have had to take had he been designing the RNA himself, his role is, essentially, a mindless one. He simply selects the ones that work best. The RNA does the R and D.

To Joyce, and all his predecessors in the field of "directed evolution," when RNA behaves like this, evolving itself, it almost feels like the RNA is "alive." Joyce's Web site banner says as much: "chemical systems express biological behavior." This is molecules behaving a bit like the living.

Could it be that the capacity to evolve is the fundamental characteristic of Life? If you return with an unsentimental eye to the list that my teacher Mr. P. Veale wrote down on that blackboard all those years ago, the first six of those "seven characteristics of Life" are significant only in that they enable a living thing to survive long enough to do the seventh item on the list: reproduce. And reproducing, you'll see if you continue to take a cold look at Life, is significant only in that it enables a species to survive long enough to evolve. If Life is an epiphenomenon of anything, it must be an epiphenomenon of the activity that appears to be the outcome of all its other activities: evolution.

It can't be just any old evolution, though. It must be a *Darwinian* evolution, an evolution powered by natural selection. In my world without skyhooks, Darwinian evolution is the only thing that comes close to the Force. Starting with nothing more than the nonliving molecules that Joyce now plays with, it has conjured up a spectacular universe of living things. I can't tell you how the trick was done, but it must have required a billion molecules to find a billion solutions to a billion problems. Along the way, the molecules associated with other molecules to form cells. Some of these cells associated with other cells to form multicellular beings. Some of these multicellular beings associated with other multicellular beings to form ecosystems. An enormous, complex, astonishing theater of function was brought into being by the Force of Darwinian evolution. And somehow, at some point during that 3.5-billion-year self-design project, Life "emerged" as the side effect of the self-design.

Now the birds in my backyard look *alive.* They flit and twitch, and peck at the soil. And they flit and twitch and peck at the soil because, within their bodies, a billion molecules are accomplishing the billion chores they've been specifically designed to do: solving their problems just as, at some point in the last 3.5 billion years, Darwinian evolution led them to. Earth alone in its solar system buzzes with the activity of all this self-design. The thing we call Life is the buzz.

Is there Life in the cultural world, the noosphere, as there is in the biosphere? It depends on what you mean by "Life." No, the noosphere doesn't store energy in the molecule adenosine triphosphate. But if, like me, you think that Life has nothing to do with adenosine triphosphate and everything to do with a capacity to "self-design" (accidentally) by engaging in Darwinian evolution, then maybe culture is alive. The cultural world certainly has its own form of evolution—cultures do change over time; you only have to visit a museum or look at what your parents were wearing in old photos to confirm this. The question is: Does this "cultural evolution" work in the same way as biological evolution? Is cultural evolution Darwinian? If it is, then culture could be regarded as alive, in a literal sense, and our new goggles will be starting to make good on their promise.

So, what's the idea? Ads and I are hightailing it through western Minnesota because I'm on a mission to answer that question. If the answer is "yes, culture does use natural selection to spout forward self-design (accidentally)," then as far as I'm concerned, Obi-Wan was right: the Force is with us always, in the biosphere and in the noosphere, and I'll be a lot closer to explaining why our species is so weird.

Mr. Darwin's Idea

Let's get one thing straight: evolution was not Darwin's idea. The idea that living things were not always as they are now, that over time they transmuted, changed from one form to another, has been around since the ancient Greeks. Darwin's unique contribution was to come up with a mechanism by which nature could have done the remarkable job of project-managing all this exquisite engineering on its own, in the absence of a designer god.

This was his idea: Every characteristic of every species has been honed by the brutality of the environment. Living things unable to boast characteristics suited to their habitat die before they can reproduce, and because they have no offspring, the suboptimal characteristics of all species vanish from Life. Nature, by being mean, designs its organisms to suit the environment, and ensures that only the best designed, the "fittest," survive.

Of course, there is one problem. The environment can change. And if the environment does change, it moves all the goal posts. What was once a good design may become average or indeed bad. And what was bad may become average or good. So nature's quality control is tireless, an ever-present pressure on living things, constantly testing every part of their design, forcing adaptation, and gradually modifying their form.

With enough time, this mindless "natural selection" could account for all the *good* design (the adaptation) and all the *new* design (the speciation) that was in evidence on Earth. There was no need for a Greater God, but there was a need for time. Natural selection is a "get rich slow scheme."[1] It has taken 3.5 billion years to get this far. When the world was considered to be only 6,000 years old, there simply wasn't enough time. It was only with the avalanche of new data from the burgeoning sciences of geology, paleontology, and archaeology at the close of the Enlightenment that we had the timeline to make evolution via natural selection seem possible.

Where did Darwin get his idea? It just so happened that at the time Time was being discovered, Darwin (accidentally*) found himself on a voyage around the ancient Earth. On that voyage he saw firsthand the evidence of the age of the Earth that others had seen (fossils of strange South American ex-beasts) and he witnessed the astonishing diversity of Life that others had reported (exotic animals and plants filling every conceivable niche wherever he looked). One landfall in particular was to become influential in his conception of natural selection: the Galapagos Islands. This is because these islands offered up something crucial to the young Darwinian mind, a body of perfect, natural evidence for the power of the Force: Galapagos finches.

Finch Mob

Although he paid little attention to the finches when on the islands (and, ultimately, had to refer to the better-catalogued collections of some of his colleagues on the *Beagle* when he got back to Britain), it was the Galapagos finch species that were to settle his mind with regard to natural selection. Indeed, given what we now know, it's difficult to imagine a

---

*The young Darwin didn't really want to join the voyage of the HMS *Beagle*. He wanted to take a short trip to the Canaries.

group of species that more enthusiastically displays natural selection's propensity for adaptation and speciation.

There are fourteen species of Galapagos finch (depending on how you split them up). They exist liberally sprinkled about the Galapagos archipelago, a cluster of eighteen volcanic islands that straddle the equator, all on their own, more than six hundred miles west of the South American mainland. The islands are young, having burst from the sea only ten million years ago, so every living thing on them can be considered a recent arrival. In that time, four different wildernesses have taken root. There is a coast of mangroves and dunes; a dry cactus scrubland; a transitional woodland with taller trees; and a humid forest of trees, shrubs, and ferns. These four habitats decorate the islands in a complex mosaic determined by the availability of water, which itself is determined by prevailing winds and altitude. The result is that the Galapagos Archipelago is in fact two archipelagoes: a cluster of oceanic islands and, upon these, a scattering of "habitat islands." Within this double archipelago the fourteen types of finches flit. They flit to the woodlands and eat seeds; they flit to the forest and dine on insects; they flit to the cactus scrub and pop the bulbous leaves; they flit to the coastal broods of giant seabirds and snack on the bugs that settle in the birds' battered feathers. The fourteen species of finch enjoy their Galapagos smorgasbord for one reason: among them, they have every type of beak you would need to grab, stab, peck, and crush the lot. Fourteen different finches, fourteen different beaks. In fact, more when you consider that on some islands the beak size of any one species can shift to such an extent that it allows the birds at the extremes to crush different seeds or snatch different insects. And more still if you consider that a single beak can be put to many uses. Take the sharp-beaked ground finch. Different populations of this species turn their beaks to a plethora of different tasks. On many of the islands they are used in the traditional way: as seed snafflers. On others they capture unwitting insects. On another they dip down into the prickly pear flower to sip nectar. On yet another they are used as a syringe, stabbing up into the skin of blue-footed boobies to steal their blood. Islands with sub-islands. Populations with subpopulations. The evolutionary kaleidoscope of Galapagos finches is dazzling.

Yet it all came from one ancestral bird: a middle-of-the-road little brown job that must have spent its days flitting, twitching, and pecking about Central and South America. The current guess is that it was the wonderfully named dull-colored grassquit, because this is the bird with the closest genetic code to the Galapagos finches. It's a good candidate for the original colonist of an underexploited double archipelago: it's not a fussy eater, happy to peck at most seeds; it lives in a wide variety of forests, from sopping wet to rustle dry; and it's used to flying over the ocean, having also conquered the Caribbean.

But getting all the way to the Galapagos Islands must have been demanding for this little bird; it's a long way. So how did it happen? Did an eager, recently mated female grassquit set off from Ecuador, as full of optimism as she was sperm, chancing her luck that there were empty islands looming over the horizon? Did a courting grassquit couple sortie too close to the cliffs and get sucked out over the big blue by a tantrum-throwing El Niño, only to be thrown down in disgust six hundred miles later? Did a little grassquit nest topple with an old tree into the Babahoyo River, sail toward the sea still hugging its clutch of eggs, and after navigating a stifling mangrove delta, make the journey westward on a playful Humboldt current, the marooned hatchlings ultimately feeding off each other to survive, screwing one another to reproduce? No, it was none of these. By looking carefully at the diversity of genes that code for the immune system in modern-day Galapagos finches, researchers have decided that the first crossing of grassquits to the Galapagos was made not by one, two, or even a clutch of grassquits, but at least thirty birds. Much to everyone's surprise, the seeding of the Galapagos finches was an act of mass immigration. But how would such a cloud of grassquits arrive at the same time? The most likely explanation, according to those who study the finches today, is that they flew out to sea when a huge fire took hold of their homelands. A katabatic wind could then have blown the smoke, and hence the grassquits, far offshore. By the time the smoke dissipated, the flock of little brown jobs would have found themselves abandoned and disorientated. Perhaps they were far enough out to spot clouds over the islands in the distance, and instinctively headed for what they thought was home. However the Adams and Eves of Darwin's finches arrived at the Galapagos, arrive they did—and from the moment they landed on the dusty soil, natural selection had fun with their grassquit form.

Two to three million years later, back home in England, lining up the delicate dead from the Galapagos trip, Darwin was staring at the results. The pressures of natural selection had forced the original grassquit beak into fourteen different shapes and sizes. There were crushing beaks, probing beaks, grasping beaks, and a beak that looked like it had come from a tiny parrot. Each beak was a tool designed for a specific job. Together they were a comprehensive set. Between fine pincers and a sturdy vise, no implement was missing. Fourteen similar bird bodies with fourteen different bird beaks—the mindless genius of self-design was laid out for Darwin right in front of the birds' faces.

Darwin felt compelled to put pen to paper: "Seeing this gradation and diversity of structure in one small, intimately related group of birds, one might really fancy that from an original paucity of birds in this archipelago, one species had been taken and modified for different ends." In these words, he described what we now call an "adaptive radiation," an outbreak of evolution that forces a glut of new species to pop up in a short time, like mushrooms on a forest floor. We now appreciate that adaptive radiations are common in nature. Whenever a new suite of opportunities opens up in the wild, a spate of new forms will follow.

My idea is to look for an adaptive radiation in the noosphere. If cultural evolution works like biological evolution, there should be the equivalent of finch mobs in the world of human thought. And where should I look to find a cultural finch mob? Why, on a cultural archipelago of course. Despite the fact that the noosphere is superphysical, it does have a geography. Since it is housed entirely within the minds of humankind, it is laid out across the planet exactly as we are. Where there are millions of us, it is deep and busy. Where there are few of us, it is thin and slow. Where we are missing, the planet is "absentminded": there is no noosphere.

If you look carefully at this mind map, you'll see that, like the double Galapagos, the noosphere is subdivided into a galaxy of "islands." The shores of each island are the limits of thought swapping, the borders of discrete cultures. In the modern world, these islands are somewhat difficult to chart—the deepest, busiest parts of the noosphere are complex places, with islands overlapping islands: double, triple, n-dimensional archipelagos of mind communities. No place for an amateur cultural naturalist like me. But in times past, the islands of the noosphere were more discernible. When we were tribal, collected in groups of two hundred or so minds, isolated from one another by unique languages and customs, the noosphere was marked out in real-space territories, just as we were. Pre-Columbus, that's what the Americas were like. They were covered with a vast, diverse archipelago of mind islands, the numerous tribes of what he would come to call Indians. Granted, many of these islands were closely associated. In the geography of the noosphere, you could say that they were clustered, perhaps even joined by spits of sand. Thoughts would have regularly migrated among islands, just as grassquits do. But the islands were not lying over the top of one another, as they are today. And these islands are easy to spot even now.

I'm driving along the I-94, annoying my brother with Springsteen on the iPod, because I'm in search of my very own Galapagos, a scattering of cultural islands that lie just beyond the horizon. Today these islands are known as the tribes of the Plains Indians. They were the last islands of Native-American thought in this country to be consumed by the tsunami of European culture. Today they don't exist as islands exactly, more as pockets of high land within the topography of general Americana. But my hope is that by visiting these ancient territories, I can map the original Plains Indian archipelago and piece together the evolution of a unique cultural article that once inhabited it, a shelter endemic to this part of the noosphere: the tepee. At one time, every tribe island in the archipelago boasted its own tepee. My task is to discover whether they were all the same "bird" or whether the tepees of the Plains Indian archipelago were in fact a genuine mob of species, an adaptive radiation of many different forms that arose as a result of natural selection in the noosphere. Who knows? In doing so, I may even distinguish the origin of tepees, that tepee "grassquit."

The first objective is to get to where the tepees once lived, the Great Plains of America, which entails a fairly respectable road trip through a series of big states. But I don't expect to be bored. If cultural evolution *is* Darwinian, there should be evidence, the telltale signs of natural selection, in the cultural world all around me en route.

Barn in the USA

I reach down and turn on the air-conditioning. It's a hot July day here in Minnesota, and I decide not simply to survive it. Passing us on either side are the farms that turned the prairie into the nation's breadbasket for many years. Each consists of a nice big plot of arable land and a focal cluster of buildings—a farmhouse, a tractor shed, one or two grain silos, a big barn—all backed by an orderly gathering of broad deciduous trees planted by ancestral humans so that descendent humans would be sheltered from the prevailing winds. It's impressively organized; a new farm appears beside the road every few miles, and in between, staggered farms are visible a constant distance from the highway. Every inch of this good soil was made use of as if in some grand plan. In less than two hundred years, human culture has turned a savannah of tallgrass and trees into a utility for humankind.

Because Ads and I are Europeans, we become captivated by the barns in particular. When we were kids we had a little Fisher Price farm set that included a barn just like these: bright red and white with a long roof laid on the barn in two pitches, so that it starts as a shallow ridge and then kinks to become a steep flank. It, too, had chimney-like structures on the top and high gable ends with big doors and a window. At the time, we assumed that it was a fantasy barn, because no barn we saw in Britain looked anything like it. Here in America, they all do.

An American barn with a gambrel roof.

Naturally enough, these barns are known as "American barns." Their special double-pitch roofs are called "gambrel" roofs, and they are a brilliant solution to a perplexing problem: they maximize the space in the hayloft without using a large volume of precious timber and without making the roof vulnerable to the continental weather. Because they are such a good design for this (physical/social/financial) environment, gambrel-roof barns are found from coast to coast on this continent, but it hasn't always been like that. The American barn didn't begin as an American barn. It has undergone an evolution, an evolution that has left its trace across the nation.

⚜   ⚜   ⚜

America was colonized by the Europeans from the East, so the farms of America are younger as you travel west. Because the farms are younger, so are the barns. Way back east, in the old colonies, ancient "English" barns remain, hidden in the valleys: proud stone or timber structures with high walls and shallow, single-pitch roofs. The main entrance to the English-style barn consists of giant opposing doors sited in the center of each long flank, so that carts can be driven in one side, loaded with hay, and then driven straight out of the other. These are just like the ones Ads and I still see driving around the patchwork quilt of Somerset in the West Country of England. Many are as old as the colonies. It makes sense. Why wouldn't English farmers arriving in the New World simply replicate their traditional barn design?

Slightly more recent (and west) are the "Dutch" barns, still found in New York State and New Jersey, where the Dutch were in abundance through the eighteenth century. They built barns out of timber with a vehicle entrance through the gable end, so the barns looked more like garages. To protect the doors from the rain, these barns had "pent" roofs that jutted out proud of the gable. Internally the barn was built like a church, with a spacious nave for the carts and an aisle on each side for livestock and storage, all supported by a dramatic H-shaped timber frame.

Farther west, in Pennsylvania, are sprinklings of "German" barns, or "bank barns," chiefly built by the German-speaking religious refugees who took up William Penn's offer to settle the colony. These are often pitched into a natural or manmade bank, so that two floors are accessible by cart: a lower stable and an upper hayloft. These were even bigger than the English barns because the German tradition was to house all livestock and farming activities in one building rather than have a different building for each, as the English did.

So it was that, armed with three different traditions of barn building from three different European cultures, waves of American farmers hurried to the western frontier at the start of the nineteenth century to grab larger parcels of land in the plainer landscape at the center of the continent. The scale of the German barn became popular because huge barns were an advantage out on the plains: the land was flat, massive, and uncontested, and each new farm was limited only by its capacity to feed livestock and shelter grain through the winter. Big barns with huge hay-lofts were fitter in this new environment, and as each barn sprouted forth from the newly ploughed prairie soil like an invasive species, the farmers, leapfrogging ever westward, gave each one a slightly bigger hayloft than the last. That is, until nature ran its course and the barns reached the point at which they could grow no further: their timber skeletons could no longer support them. The European barn-building traditions had run out of "design space."

But then a radical mutation swept through the population. Barns were being born with ingenious double-pitched roofs. The roof trusses, when built in the two pitches, supported themselves, so the barns had no need for any internal posts, and could have a voluminous hall for a hayloft. They were even better than the European barns. No one knew what to call them. "Gambrel roof" became the nickname, some say because the profile of the roof looked like a gambrel, the joint above the hoof of a horse's rear leg.

Out of nowhere, a spontaneous new barn—a brilliant solution. The first in the New World to be named not after an old tradition from an old nation, but after a brand-new one. This was the *American* barn. Once born, its design rapidly spread both westward and back eastward. Bigger, better, American, and, most important, not European, the gambrel-roof barn represented in physical form all the hopes the Americans had for their New World. It was a whole hunk of timber-built progress and it became an icon of a young nation. Which is why Ads and I came to be playing with a plastic version of it back in Europe in the 1970s.

But Who invented the gambrel roof? Was it the work of an immortal—did God *bless* America with this kinky roof? Was it the work of a mere mortal, a farmer/design hero of the Midwest? Or was it a prime example of the sort of (accidental) self-design that comes about if culture is powered by the Force of Darwinian evolution?

4

Variation, North Dakota

Plains Sailing

Entering North Dakota at Fargo, we trundle over the Red River of the North and shuffle through the stoplights of downtown. Then, after a flurry of gas stations, fast-food joints, and motels, Fargo ends and we find ourselves back in the great outdoors. But it's changed. In place of lush, rolling Minnesota farmland, a huge horizon confronts us—flat, brown, and empty. North Dakota looks like it's been ironed. It's the start of the Great Plains, a landscape so flat the Fargo Golf Club had to build hills to make its course more interesting; the pins on the seventh and eighth holes are the highest points in view.

Our bearing is now due west. Our plan is to head clear across the state, a distance of more than three hundred miles. Leaving humanity in Fargo makes the prospect quite exciting. Waves of grass flow away on either side of us. The land begins to rise and fall, so, as we skim along at sixty knots, the car gently pitches up and down. We're driving on the Great Plains, and it feels like a sea voyage.

In a sense this is literally true. The land here rolls *gently* because, for millions of years, it was the bottom of an ocean called the Western Interior Seaway, which wallowed in the belly of North America until it dried up with the dinosaurs in the late Cretaceous. Since that time, this old seabed has hosted everything from deserts to frozen rivers, but when, ten thousand years ago, the ice retreated for the last time, the land was submerged by a new kind of sea: a sea of grass over half a million square miles in extent.

The approximate expanse of the Great Plains.

Plants divvy up land democratically. Where water is delivered, by clouds or rivers, faster than the sun and the wind can evaporate it or the cold can freeze it, trees grow. Where water is delivered at a rate that is less than half as fast as the sun, the wind, and the cold can remove it, there is desert. Everywhere else, you get grass.* For ten millennia (barring a few local shortages), the Rocky Mountains have allowed just enough water through to the plains to deny a desert, but not enough to start a forest. It was the creation of the Rockies in the late Cretaceous that first drained the Western Interior Seaway to form the plains, and sixty-five million years later, the mountains still run the place by choosing how much water gets through to the heart of the continent from the sodden Pacific Coast. As a consequence, this immense grassland hangs off the eastern front of the Rockies, all along its march south across the continent, like a devoted pet dog walking perfectly to heel.

To the untrained eye, this "wilderness" may look uninspiring. But grass is not just *grass*. There are hundreds of different species of it, each with its own techniques for denying the sun and/or wind and/or cold their precious water. They vary in height, width, growing season, flowering season, root depth, dependence on different minerals, tolerance to waterlogging, number of spiracles on the leaf surface, and any number of other characteristics, in order that there will be at least one patch in the bewildering space of the grasslands that suits them above all others. The grasses in Alberta can outwit the freeze in winter; the grasses in Texas take on the summer sun with a smile. Those in the West have a mineral appetite perfectly suited to Rockies rock, which is served daily, in molecule-size portions, by the eager, eastward-flowing rivers that drop down from the mountain valleys. Grasses in the eastern part of the plains dig their roots deep into the hills and feast on the antique silt blown down from the melting glaciers at the end of the last Ice Age.

The old Western Interior Seabed tips slightly, so that as we travel west, toward the mountains, we'll be moving gradually uphill and into colder climes. In addition, the rain shadow of the mountains is more effectual in the West, so that as well as being higher land, it is drier land, and the grasses must settle themselves according to their powers of water retention. This gradient from east to west—ever drier, ever cooler, ever higher—paints three vertical stripes across the Great Plains, if you happen to look at them in an agricultural science textbook. Where we now are, on the east of the plains, flamboyant, lanky grasses flap their broad leaves about, enjoying the deep "loess" soils and the best of the rain.

*ish . . .

They shoot up to seven feet in a single growing season before dropping back to take on the winter. This is the "tallgrass prairie." Farthest west, on the high plains and under the mountains, the grasses are short, with narrow blades and deep roots. They put up with the hard life, clinging to any water they can save, growing achingly slowly, shutting down when no water comes at all. This is the "shortgrass prairie." Between the two lies the "mixed prairie," a meeting point for grasses from both east and west, where the eastern tallgrass must suffer the humiliation of stumpiness and where the western shortgrass can stretch.

Of course these bands are not visible at ground level; there isn't a spot in Nebraska where the grass community suddenly gains a foot and a half. There is a gradual admixture running east to west so that you barely notice the change as you drive alongside it. A businessman journeying east from Bozeman, Montana, toward us in North Dakota wouldn't notice the grasses change; he would just get the impression that his tires were losing air.

The "mixed prairie" officially starts around the middle of North Dakota. It's nightfall before we get to it, and as I see the lights gather ahead, "Bismarck," I say to my brother, "the capital of North Dakota." I try to put gravitas into the announcement, but my stage voice fails as the same gas stations, fast-food joints, and motel chains return. We wade in and find a motel.

Barn Different

Natural selection is a recipe with three ingredients: variation, inheritance, and selection. If you have all three, you can't really get the dish wrong, no matter in what proportion you add them, nature will cook you up something good. The American barn, had it been a product of Darwinian evolution, would have stemmed from such a recipe. But can I detect these ingredients in the barns I see? Let's start with the first on the list: variation.

After a regulation motel breakfast, Ads takes the wheel, and we head north on a little day trip. Now that I don't have to watch the road, and now that the barns are so much closer to the verge of this small highway, I have the opportunity to spot the subtle variations in their design. While most have straight, if double-pitched, roofs, some have rounded roofs, shaped like the arch in a cathedral doorway or a conquistador's helmet.

While most have roofs that stop halfway to the ground, some have roofs that go almost all the way down. Some have blank timber gable ends, while some have small square windows near the top, like cute little eyes. Some have two big doors that swing on hinges. Others have a single, great sliding door. A few have little "chimneys" (called cupolas) on top, which admit light and air into the hayloft. Some have roofs that extend out at the bottom to create a lean-to at the barns' flanks.

It seems a facetious question to ask, but why do they vary? The answer we'd all come up with (after a "tut") is "because the local farmers wanted them to vary, of course; they built bespoke barns for their particular needs." But I'm a cultural naturalist now, trying to discover more about the apparent Life of human culture, so I have to tease this further.

In Darwin's day there was a craze for collecting large numbers of individuals of one species in order to chart Mother Nature's "natural variance." Rich Englishmen with nothing better to do would go out butterfly catching with big floppy nets and bag bundles of "red admirals" and "cabbage whites." Back at home, in their studies, they would pin them to specimen boards in series that best displayed the gradients of difference they had found in the natural population, perhaps a growing blotch of yellow on the fore wing or longer and longer tails on the hind wing. This seemingly harmless pastime was in fact to fashion a significant leak in the levee of creationism. Over cigars, the Victorian butterfly enthusiasts were destined to ask, "Why do they vary?" When the answer came back (after a "tut") "Because God wanted them to vary, of course; he built bespoke butterflies for his particular needs," it simply prompted other questions, such as "Why wouldn't he make only perfect butterflies?" and "What particular needs?" The furrows on the Victorians' brows deepened as they puffed, and the resulting awkward silences set the stage for those who would bring a revolution in our understanding.

To paraphrase, here's what those revolutionaries concluded: Nature doesn't make "perfect." Instead she mass-produces "about right," just like the manufacturers of tuna melt panini. There is no "perfect" because the environment is always changing in subtle ways, tweaking the selection pressures, killing each generation off differently from the one before. The in-built imperfection of each individual is actually beneficial. It works like an insurance policy. Regardless of how the environment changes, the theory is that, among the variants, there should be at least some individuals that can deal with the new regime of "particular needs." If shorter tails are called for in the advent of higher winds, then we have some of those. If bigger blotches assist the advertisement of a butterfly's warning colors in the advent of duller days, then we can cope with that change. Although Darwin didn't come up with the idea of natural variance either, it was a key component in what was to become known as Darwinian evolution, and (if I squint) I can see it here in the barns of Minnesota. They vary like butterflies: none is perfect; each is slightly different.

But do barns vary *like* or *exactly like* butterflies? Is the similarity in their variation a coincidence, or is it due to an underlying common causative factor, the action of Darwinism? There's a straightforward objection to the proposal that butterfly and barn variation have a common causative factor: butterflies vary because every individual has a slightly different genetic code, which is to say that every individual has a slightly different set of instructions on how to build a butterfly. Hence each butterfly egg hatches and each butterfly is built slightly differently as a result. Barns don't have a genetic code. So, where do their slight differences stem from?

In the days before industrialization, no farmer could build something as massive as a prairie barn alone. Instead, there was a tradition of barn raising, in which the entire community descended upon a new farm and, from nothing more than timbers, shingles, and handmade pins, constructed a building in a day or two. In return, the workers would be fed with the best produce and beer on offer in a feast that celebrated not only the continuation of barn craft and tradition but also the minor victory of taming another patch of wild America. These barns were not designed by any one person. They arose from collaboration, and this accounts for much of their "natural variance."

Picture a barn-raising team coming to build "Olafsen's" barn: dozens of men and women, some with years of experience, others without. Plenty of them will have strong opinions on what to do when. As the barn goes up, these opinions will be voiced. Discussions will ensue. Decisions will be made. The barn will materialize only upon the successful cooperation of scores of barn-building instructions from dozens of barn-building minds. Eventually, after completing its own personal negotiation with the communal repository, Olafsen's barn is finished.

The following spring, it's "Larsen's" turn. Another barn-raising team congregates. But there are slight differences in the personnel. The building of Larsen's barn is a novel collaboration, and since it's a novel collaboration, it will be a novel creation, made from a slightly different string of instructions. Even if they want to, it will be impossible for this team of barn raisers to raise a barn identical to the last. In fact, *even if exactly the same team turned up*, the discussions and decisions that led to Olafsen's barn would not be replayed *exactly*. The mixing of Dennettian thought-swapping technology with Popperian problem-solving technology must surely make a clone barn extremely unlikely. And anyway, even if exactly the same team turned up, *and* the discussions and decisions were replayed exactly (by some incredible fluke), Larsen's barn would *still* be different from Olafsen's, due to another factor: the influence of an ever-changing (physical/social/financial) environment.

The constantly wavering real world will inevitably make its mark on Larsen's barn. Perhaps, that spring, there aren't enough long timbers available and the barn raisers have to use more joints than they normally would. Perhaps, on that site, there is a slight incline in the land, and the barn raisers have to tilt the foundations a tad. Perhaps, on that occasion, a boyish rivalry between two sets of brothers begins a nail-bashing race, and the whole roof ends up taller than it should have. The varying environment leads to varying barns. It ensures that even well-considered, well-rehearsed barn designs are realized *imperfectly*.

Olafsen's and Larsen's barns, and all the others that glide past us as we drive north of Bismarck, were destined to be born different. They may not have a genetic code, but they do vary exactly like the butterflies of Victorian gentlemen: as a result of a never-to-be-repeated interplay between a competing community of instructions, the peculiarities of the local environment, and a pinch of good old-fashioned randomness. Because the barns were built by a team of collaborators working in the real world, they couldn't be clones. American barns bred their own natural variance.

That's not the way it works today, though. For two centuries, all across America, barn-raising teams were raising unique barns, taming the wild with their imperfect toil. There was one barn every minute! But with the close of the nineteenth century, the tradition fell prey to mass manufacture, and the marvellous American barn could be bought by mail order, arriving ready-to-build from far away. Concrete floors and machine-cut timber standardized their creation, and the barn lost its natural variance. The vast twentieth-century barn factories *were* able to make barn clones—indeed, that was their purpose. Barn-building savvy left the local communities (the amateurs destined to make each barn slightly different) and harbored only among the neurons of a clique of professional engineers, with their own particular needs: namely, to maximize profit.

As we dip off the highway and onto a small rural road, the largest grain silos I have ever seen come into view, as if to emphasize the industrialization of agriculture. Two lines of aluminum skyscrapers rise from the fields, bursting with pipes like those on an oil refinery. They even have their own railway station to ship out country produce to the famished city folk.

Staring back at these monsters, we nearly miss our turn. We make a right and drive through still, silent Stanton. We're near this morning's destination, and to announce that, Stanton is decorated with silhouettes of Sacagawea, the Native American woman who accompanied the 1804–1806 Corps of Discovery expedition led by Meriwether Lewis and William Clark. She lived hereabouts. A few miles up the road, we discover exactly where.

A Port on the Plains

Perched on a mud cliff, our backs to the Knife River, Ads and I, baking in the ninety-five-degree temperature, squint at a broad, recently mown meadow. The high sun cancels all shadows, but we can still easily see a densely packed array of large bowls in the field. There must be more than forty of them, and each is thirty or more feet in diameter. They are the dents left in the ground by earth lodges, the summer homes of Native Americans who lived in this valley until the middle of the nineteenth century. The earth lodge village that existed here was one of a number in the area built by three unrelated but mutually respectful tribes: the Mandan, Hidatsa, and Arikara. The oral histories of these people suggest that they all originated in the eastern woodlands and came west along the river system around five hundred years ago, pushing deep into the Great Plains on the Upper Missouri, until even the river valley gave up hope of hosting trees. Here, where the woodlands ran out, they settled, tree people clinging to the very edge of their realm, the Great Plains looming all around them.

In the East they had learned the skill of farming, a set of instructions that had ultimately filtered northward from the Aztecs in Mexico, the first accomplished farmers in the Americas. The three tribes knew how to raise corn, beans, squash, sunflowers, and tobacco. They had placed their gardens on the eastern floodplains, where the soil was rich. It was valuable land, and they had to protect it. Fortunately farming could yield lots of food from even a small patch of land, so they didn't need to maintain the huge territories of their hunter-gatherer ancestors, and the bounty of their harvests naturally led to dense populations. Each village was well defended by numbers.

Ultimately, though, as the skill of farming spread among the eastern tribes, the East began to fill up, battles became more frequent, and the three unrelated tribes independently sought peace by fleeing west along the rivers. The rivers took them not only west, but north. Eventually they met up in the Upper Missouri.

This new homeland would test their native culture to the extreme. The habitat they recognized was hair-thin, snaking through a landscape that must have appeared as bare as the moon. Getting a good crop was hard because of the short, unpredictable summers. The tribes supplemented their disappointing yields by promoting the importance of another skill: fishing. The cold rivers had plenty of good, fat fish, and the tribes would set traps for them in the lea of sandbars by placing coiled willow panels in the riverbed and baiting them with maggoty meat tied to a cottonwood sapling. The old men of the Bear Society would sing fish songs and smoke tobacco until the sapling rustled and they knew their supper was caught.

But the fish were not enough to survive on. The tribes had no choice but to explore the enormous, empty world above their safe valleys. Wandering out onto the plains for the first time must have been terrifying— the landscape is endless and featureless—but the endeavor brought an unexpected reward. They found that if you knew where and when to look, the empty plains would suddenly fill with the greatest meat source on the planet: colossal herds of plains buffalo.

But the plains buffalo was an alien species to these tribes. To feed on them, the Mandan, Hidatsa, and Arikara were forced to change their way of life. They had to adapt to become an alien species of Indian: a "Plains Indian."

The Plains Indian was the noosphere's answer to the question "How can Stone Age humans live on the Great Plains?" Theirs was a brutal and flamboyant existence. We all have images of it from a lifetime of cowboy-and-Indian films: the headdresses, the peace pipes, the "powerful medicine," the taking of scalps, and that "sun dance" that nearly killed a man called Horse. But these were just decorations. At its heart, Plains Indian culture was a nomadic, hunter-gatherer strategy that came into being only upon the adoption of three requisite subcultures.

First, there was the culture of buffalo hunting. Buffalo are huge, jumpy animals with a thousand pairs of eyes, a thousand pairs of ears, and a thousand noses (due to their keenness for living in groups). They frighten easily, and when frightened, they move as one in a stampede that shakes prairie dogs out of their burrows. Hunting them is no picnic, but the Plains Indians devised a hundred ways to do it, and could choose the perfect way to suit the prevailing topography, weather, and size of herd. They even knew ways of killing buffalo without even touching them. They could butcher their quarry on the spot, working together in a factory line, ensuring that no part of any animal was wasted: they would boil up the bones to extract a glue; they would sew together the bladder to make a water canteen; they would even use the buffalo dung that remained in the carcass's gut to fuel their celebratory fires.

Second, there was the subculture of pemmican, a high-energy foodstuff made from dried and pounded meat, melted animal fat, and, often, dried fruits such as choke cherries and Saskatoon berries. The tribes would make pemmican when the hunting was good in summer and fall, then wrap it tightly in rawhide pouches and bury it in caches for the winter and early spring. The European trappers who first traveled on the plains quickly understood pemmican's importance. Without it, it would have been impossible for them to survive the Great Plains winter.

Finally, there was the tepee. As a well-known song declares, "the buffalo roam," so if your life depends on the buffalo, you must roam, too. And that means having a home that can roam. For the Plains Indians, the tepee (or teepee or tipi) was the answer.

"Home Sweet Home"

The tepee is more than just a tall tent; it is a marvel of human ingenuity. With its self-supporting skeleton of lodge poles leaned against one another, the tepee boasts a completely unobstructed interior volume, so that its inhabitants can walk around upright and unimpaired. At the "nest" high above where the lodge poles intersect, a natural smoke hole allows the soot and smoke from a central fire to escape without ruining the eyes of the family huddled around it. To assist the smoke upward, the tepee has a lining that directs fresh air from the footings up to the smoke hole ventilating the interior. The lining is also an insulator, keeping out both the harshest winter drafts and the hardest summer rays: like the Mall of America, this Home Sweet Home can maintain the perfect temperature for its inhabitants' cave bodies.

As well as being comfortable, the tepee is eminently practical. Its wrapping is made entirely from the most abundant sheet material on the plains: buffalo hide. Its conical shape may stand proud of the plains, but it's virtually impossible to blow a tepee over. It can be put up or taken down in less than half an hour. And once it is dismantled, its pine poles can be arranged in such a way that the tepee magically turns into a sledge large enough to carry itself, most of its former contents, and a few children to boot.

To the Plains Indian, the tepee was more than just a house; it was the difference between life and death in this, one of the world's least hospitable wildernesses. It enabled these grass sea gypsies to sail freely over their ocean, to fish wherever the shoals of buffalo were thick. For the fully nomadic tribespeople, those who spent even the deep winter up on the turf, it was the only home they would ever know.

The Mandan, Hidatsa, and Arikara never committed to the Plains Indian way of life to that extent. They never abandoned their gardens, and they never fully left their earth lodge villages. They used their tepees only when on their short buffalo hunting trips, sailing out from their ports on the edge of the plains, never far from home, hunting and butchering small numbers of buffalo, dragging the carcasses down to prearranged loading spots on the smaller rivers, where women in round "bullboats" would meet them and cart the goods back to headquarters.

So it was that Lewis and Clark found a huge earth lodge village of Mandan and Hidatsa on this site when they got to the Knife River in October 1804. The waterways were full of bullboats, the village a haven of busy humanity in the region. Since these natives were the only non-nomads near the shortgrass, they had become by default the chief traders among the Plains Indian tribes. They had access to rare quarries of flint, and served as the middlemen for the movement of many goods throughout the region. Lewis and Clark and their Corps of Discovery had found the perfect people to help them get farther west: the three tribes were the only boat people in the area, and they knew the paths of all the rivers right up to the Rockies. Plus they happened to have a key asset among them: a native woman called Sacagawea, whom they had kidnapped as a girl from the Northern Shoshone tribe, the tribe that lived on the other side of the Great Plains, at the headwaters of the Missouri River. She still spoke her native tongue and still knew how to communicate with neighboring tribes using a Plains Indian sign language. She was instrumental in securing horses for the Corps on their journey and leading them through the Rockies and over the Continental Divide. Lewis and Clark returned to St. Louis in 1806 with only one fatality. Sacagawea was responsible for much of their success.

At the site museum, Ads and I find a reconstruction of the sort of building she and all the Mandan, Hidatsa, and Arikara lived in. The earth lodge is a huge circular mound of soil, approximately thirty feet in diameter, with a high wooden tunnel entrance on one side. We are instinctively drawn in. The day is unbearably hot, but in the replica earth lodge the cool is deep and satisfying. We each sigh in relief as we escape the sun. The tunnel ends blindly ahead, and we are channeled through a huge buffalo skin door to the right. As soon as we are inside, we understand the asymmetrical entrance: the center of the lodge is given over to a fire. Above it, a large central ceiling hole allows the prairie summer sun restricted access to the room. At this time of day, sunlight charges in and slaps a wall of cut trunks, the reverse side of the entrance tunnel. This wall is coated with a splayed animal hide, and the hide glows with the yellow of the sun. Painted across the hide from its center are rings of rays and symbols: a reconstruction of the robe of the earth lodge's owner.

It takes a moment for our eyes to notice the room that runs around this brilliant, blinding fire pit. There are a series of beds raised off the floor, each surrounded by a privacy canopy made from more buffalo hides. There are axes and knives, balls and sticks for playing games, a makeup kit, a simple shrine, pots, and rawhide bags. We're prevented from freely wandering by a line of museum tape, but just beyond it we see a dugout cache pit: the earth lodge refrigerator.

It is a magnificent, roomy, comfortable home, and a credit to the people who developed its design. But it isn't unique. Shelters like this were also used by the mound-building Mississippian cultures of what is today Illinois and Kentucky, by the Apache and Navajo of the southwestern deserts, and by the natives of British Columbia. While the three tribes of the Upper Missouri certainly perfected the earth lodge, they probably didn't invent it.

But what about the tepee? Were the Mandan, Hidatsa, and Arikara the first to use this remarkable structure? Did the tepee originate during a spectacular brainstorm held in one of their earth lodges?

No. The truth is that the Mandan, Hidatsa, and Arikara were part-time Plains Indians. They weren't committed enough to the plains to have developed an entire Plains Indian culture. With the Omaha, Pawnee, Oto, and Wichita to the south, and the Ute, Shoshone, Nez Perce, and Yakima to the west, the Mandan, Hidatsa, and Arikara formed a ring of outer tribe islands, a periphery that marked the edge of both the plains and the tepee radiation. These tribes were all seminomadic: part-time tepee dwellers. They would dip into the plains buffalo stocks in summer and retreat to the fringing savannah, woodlands, or mountains when the bad weather came. To have a hope of finding the origin of tepees, I need to search out the full-time nomads, the people who spent their entire lives on the shortgrass. The people who could not be without a tepee. It was they who had the necessity to mother that invention.

At Knife River, the only tepees on display are three bright white canvas mock-ups that stand in a circle outside the museum. They don't have any linings, and their doorways face in different directions; I'd always read that tepee doors faced east. "We made those with some local school kids on our Cultural Day last week," says the helpful Scandinavian American assistant at the museum. That explains it.

5

Inheritance, South Dakota

Biological Brothers, Cultural Cousins

We retrace our steps to Bismarck and turn west. Less than an hour later we realize that before, when we felt we knew what the real plains were like, we were mistaken. We've arrived at the shortgrass, the high, dry, hard, greasy grass that at one time served as a buffalo thorough-fare. If the plains are a sea of grass, then we've just left the continental shelf: this is the open ocean. Gone are the trees, the lakes, and those occasional pillow hills. Ahead is a landscape so vast and empty that time disappears into it. The Chrysler cruises for an hour and a half before I have to move my hand on the wheel. There is literally nothing to see here. The only décor is the occasional road sign: the comforting "Rest

Area 91 Miles," the assertive "Be Nice," and the prudent "Fear God, Not Terrorists."

Dickinson, North Dakota, interrupts my lack of concentration. It comes out of nowhere. It is nowhere. Speeding billboards advertise a wurst shop and a smorgasbord. But we don't stop to sample. We cruise on, ever westward, until, thirty minutes later, we're sent for the first time into a violent bend, and the earth suddenly gives way. Bump! The plains disappear before us. In their place is a maze of sand hills that seem to be painted in layers of yellows and pinks. Transported abruptly from such a stark landscape to one so hectic with detail, our brains do the equivalent of shunning a bright light. It's only when my mind adjusts that I can make sense of this new view. Some hidden river must have eaten a chunk out of the plains, revealing the rainbow of layered history that lies under the grass. We've found the northern chapter of the Dakota Badlands, and as tourists, we don't react well. We shout and make a sudden movement to pull off the highway, eager to have something to look at. Thankfully the rangers at the Theodore Roosevelt National Park have anticipated this. They offer us a car park, a viewpoint, the rest room that was prom-ised us hours ago, and some chilled water. We gaze and sit, sipping, on a fence overlooking the vast painted canyon.

An hour later, Ads and I have done the tour, gotten the T-shirt, driven the scenic route. We've taken short walks to look over the sweeps of the impatient "Little Missouri." We've enjoyed our first sighting of buffalo, prairie dogs, and mustangs. Ads liked the mustangs best. They were splattered with brown, blue, and black, a feral hodgepodge, just like they ought to be. We could picture the old Indians riding them. I liked the buffalo, lying by the side of the road like disgruntled cows, with cartoon heads and horns, swishing their flies, oblivious to the history they'd been involved in. The bulls are so heavy at the front you feel they will tip, their silly little back legs left to kick fruitlessly in the air. But they never do. Perhaps their skulls are counterbalanced by the weight of their low-hanging and enormous bollocks.

Satiated with Great Plains fauna, we drive to the crossroads with Highway 85, which heads south, in awkward steps, to Deadwood, South Dakota. "D'you know, only one of those two, horses and buffalo, is an authentic Great Plains animal," I pronounce to Adam. "Guess which." "That's easy," he says. But it isn't; I'll tell him why later.

We pull up at a gas station.

Inside, sharp pairs of eyes stare at us, baby blues peeking out between smears of oil and mud. The nice, neat farmers we'd occasionally spied among the tallgrass are no more. These are the grubby farmers of the shortgrass: shorter, darker, wilder, more desperate. Stained baseball caps sit low over their brows. There are handlebar moustaches everywhere, and I've not seen those since we landed. The shortgrass was never meant to support agriculture, and these guys now know it. "Farmers east of here are all millionaires," our waitress in Bismarck had said, "but west of here—that's a hard place to farm."

"D'you think they think we're gay?" I mutter to Ads, as he looks for the vegetarian option on a stand of do-it-yourself nachos.

"No, we look like brothers," he replies.

I laugh as he covers his chips with a "chili sauce" that turns out to be chili con carne. Then realize that *I'll* have to eat that now.

Ads turned vegetarian a few years ago. It fitted with his approach to life. He likes the mystical. He likes the idea that there is more to the universe than I can allow myself to believe. He's a vitalist. He uses his chi all the time, even before breakfast. World of Warcraft, Metallica, and, now, shamanism followed the vegetarianism. It all seems to fit, but I do sometimes ask myself how we could be so different. We had the same parents and grew up in the same house. What is it that led us to think in such different ways?

It's all to do with the laws of inheritance as they apply to our weird species. Because we are humans, creatures that inhabit two worlds, Ads and I, sitting side by side in the front of our Chrysler, are subject to a peculiar dual inheritance. Our biological inheritance, the heredity of our *biospherical* selves, is almost identical. The DNA in his and my old skin cells—settling in a muddled dust on the clutter of semi-disposable drinks cartons and chips packets on the backseat, sloughed off each time we remove our fleeces as the sun climbs in the morning—contain nearly the same genetic code. But our cultural inheritance, the heredity of our *noospherical* selves, is very different. For at least twenty years, we've led different cultural lives. I know plenty of things that he doesn't. He knows plenty of things that I don't. And we "value" the different things we know in different ways. The result is that while we are like-bodied, we're not that like-minded. We may be brothers in the biosphere, but in the noosphere we are only distant cousins. Our respective databases of cultural information are dissimilar, and getting more dissimilar each day, because, unlike biological inheritance, our cultural inheritance continues until we're dead.

"What's that?" says Ads, to prove my point.

"A pronghorn antelope," I answer.

We notice that antelopes are popping up on the side of the road as often as moustachioed, grubby farmers during our long drive south to Deadwood. I suggest to Ads that perhaps they are both creatures of the shortgrass.

Now that we're cutting through the Great Plains at right angles, we can't ride the topographical swell anymore, and we're abruptly sent up and down over huge grass-ocean rollers. As the land heaves, a distant horizon repeatedly appears and then disappears. It doesn't take more than a glance at the map to realize how far we have to go before nightfall. We want to get to Deadwood, and we don't want to pitch our tent in the dark. The temptation, in this emptiness, is to floor the gas; however fast we drive, it doesn't seem to matter to the plains. But after only a few miles, we suddenly spy a highway patrol car parked in anticipation on the verge, and Ads tinkers with the brakes, trying to slow down without looking like he's trying to slow down. We approach the cop cautiously, expecting trouble, but then discover that the town of Amidon has a sense of humor: a mannequin in shades and a sheriff's uniform slumps back in the driver's seat.

Getting back up to speed, flying past farms, we start to spy the American barns of the shortgrass. They've clearly had a harder life than the ones back east. Many are leaning precariously on their knees. They all look old, and because they all look old, it's difficult to work out exactly how old they are. The barns of the 1920s resemble the barns of the 1890s, which resemble the barns of the 1870s. And I realize that right there, in their repeated resemblance, is proof that culture must be inherited. The barns resemble one another for the same reason that Ads and I, and Ads and our father, and our great grandfather and I resemble one another: we have a common ancestry, a shared inheritance.

The Front of the Barn

Inheritance is the second ingredient of natural selection, and it's another idea that Darwin didn't come up with. Heredity, the passing on of traits from parent to offspring, has been regarded as a fundamental feature of nature since ancient times. But it is only in the light of Darwin's work that we came to appreciate how important it is. Without the capacity to inherit, Life would never have gotten going, because inheritance is the solution to Life's biggest problem: death.

In the end, all living things die. This is so true that Mr. P. Veale could have used "all living things die" as one of his defining characteristics of Life, and "dying thing" would be a suitable alternative to the phrase "living thing." It is of course tragic, but it's also catastrophic, when you consider that with each death a whole lifetime of experience on how to survive on planet Earth also disappears. And if the experiences of all living things come to nothing, then Life can never move on, grow, "progress." Life had to come up with a method by which one living thing could pass on information—instructions on how to live successfully on Earth—to the living things that came after it. About 3.5 billion years ago, it did.

A molecule came together (accidentally) that worked a bit like a magnetic message board on a fridge; it could hold on to instructions so that other organisms could read them later. The molecule was DNA, and the different instructions it held on to were genes. Ever since then, all of Life has used genes to transmit information to the next generation, but instead of sticking them on the fridge, the "parents" neatly fold up copies of their instructions within the bodies of their "offspring," for them to (chemically) read later. In information terms, it's like cheating death: the individual living thing dies, but a copy of its essence comes to life again in another. This *offspringing*, this cheating of death by copying, is the fundamental pastime of Life.

Does the noosphere have a system of inheritance? Of course it does—the copycat barns I'm seeing are evidence of that—but what exactly is going on in this form of inheritance? How can the barn door on one barn be almost identical to the barn door on a barn built sixty years earlier? What is it that is being copied through the generations?

It pays to scrutinize biological inheritance more closely. Looking at me and my mum, you might say that I've got her nose. In fact, I haven't; she still has it. What I have done is inherited copies of the instructions (genes) that she has for "nose design." The nose is just a front; I have copies of the things that know how to make the nose. It's exactly the same with the barns. The barn doors, the cupolas, the trusses—they are all just a front. In fact, *the whole barn is just a front*. The things that are inherited down through the generations are the copies of the things behind the barns, the instructions for barn design. And where do these instructions reside? Not in the barns at all, but in the minds of the barn raisers. In place of gene inheritance, the noosphere has *thought inheritance*, and for creatures like us, the crème de la crème of Dennettians, it's the most natural thing in the world. Look, I'll demonstrate it right now.

Imagine that spring has finally sprung, the blossoms are out, and the days are turning warm. You decide that today's the day you're going to start playing outdoor games again. You rush to the garage and set up the foldable Ping-Pong table, rig the net, and get the Ping-Pong paddles. But you can't find the balls. You look behind the bikes and gardening tools. Eventually you find one ball, but your joy is short-lived as you rotate it to see one of those pursed dents that most Ping-Pong balls have. Someone trod on it last year.

You can't be bothered to go down to the store for one Ping-Pong ball, so you give up, de-rig the net, refold the table, pick up the paddles, and push them all back into the garage, and because you're cross, you stub your toe on the way just to make the whole experience even more unsatisfying. That is, unless you know a way to undent a Ping-Pong ball. You see, there are two types of people in this world: those who know the Ping-Pong-ball-undent trick (and who go on to play a fun game of Ping-Pong with their family while the cherry blossoms land all about and the bonus sun shines), and those who don't (who waste half an hour and hurt their toe).

Which group do you belong to? If you're in the former, I can't do anything with you; you have already used the noosphere's thought inheritance system in this respect.* If you're in the latter group, then don't worry; you're just about to. Here's how you will avoid some agony next spring: First, jog into the kitchen, boil up some water, then drop the dented ball in and wait for that reassuring but faint "pop" sound. Bingo! Usable Ping-Pong ball. The hot water heats up the air inside the ball so that it expands and pushes out all but the most cantankerous dents.

This is something you may have worked out yourself if you pondered

---

*In fact, you'll no doubt be thinking, "What a lame example. Everybody knows how to undent a Ping Pong ball." But I can tell you that they don't. I was thirty-eight when I learned this trick. I'd been playing Ping-Pong for over twenty years, ignorant of it. Now I have that "that's easy" feeling like you do. That's the sort of feeling you get when you're cultured.

it long enough. But we rarely do that, do we? The Ping-Pong-ball-undent trick is almost always *learned* rather than *invented*. To humans at least, learning a solution to a problem (the Dennettian way) is far easier than inventing a solution to a problem (the Popperian way).

If the Ping-Pong-ball-undent trick was a revelation to you, then, congratulations, you have just inherited an instruction on how to cope with a particular part of life on Earth without a single gene changing place. You've inherited a thought that was in my brain. Psychologically, this is a remarkably complex thing to do, but it felt easy, didn't it? Well, it would, since thought inheritance is what we humans, the crème de la crème of Dennettian creatures, the most cultured creatures in the world, are built to do. Thought inheritance is our USP.

"Pronghorn," says Ads. (See.) "Pronghorn," says Ads. "Pronghorn," he says.

Dead Man's Hand

For the first time today we spy clouds ahead, coated in gold because the sun is about to change shifts with the moon. As all mariners know, clouds give away the location of islands, and this holds true here. The big rollers of northern South Dakota come to an end, and we start to drop in altitude. As we tip forward, the anticipated island comes into view: a dome of dark trees and craggy cliffs welling up from the plains. It's the Black Hills, a geological anomaly on this peneplain. Isolated, surrounded on all sides by millions of acres of shortgrass, it couldn't look more different from the landscape we've been sailing on. There is a sense of relief in the car. It's been an odyssey across the blades from Bismarck, but with our destination in sight, we tack the Chrysler around and make for it.

In no time at all we are at its shores. The car begins to snake upward, and we become enveloped in ponderosa pine forest. It's cool and lush. We drop the windows to suck in the wet air and the forest scent. Our faces lose their plains dust to the damp of the trees. Lakes appear, and horse farms in the valleys. It's like another world. Within a few miles, we are high among the crags. We enter the town of Deadwood.

Deadwood is a remarkable place. At the start of 1876 it didn't exist. By the end of 1876 it had five thousand inhabitants, a main street with more than seventy saloons, and its own newspaper. What had the power to force the development of such a community in a part of the country where white America had never settled? Gold. It was the ostentatious 1874 expedition of Lt. Col. George Armstrong Custer that found gold in them there Black Hills. Custer, already a Yankee Civil War hero, was sent to scout out the sacred mountains of the Sioux in order to find a good site for a fort "from which those hostile and disgruntled indians might be controlled." This was unexplored plains country, so Custer took no chances. He had more than 1,000 men in 110 wagons. He had companies of the Seventh Cavalry on his flanks. He had a brass band that played rousing tunes throughout the expedition. He had journalists, a photographer, scientists, and prospectors. The entire ensemble had its own particular marching order. It must have been a surreal caravan, out on the shortgrass, taking nearly three weeks to cover exactly the same route that Ads and I have just done in a day, from Fort Abraham Lincoln (Bismarck) to the northern edge of the hills. But it did what it was supposed to, and much more: it found gold, and consequently founded Deadwood.

Gold rush Deadwood was born rough as hell. Opium, guns, hookers, drinking, and gambling—peddled by saloon owners, paid for in gold dust. Lower Main Street was particularly grimy; everyone referred to it as "the Badlands." Many ended their days here, but of the ninety-seven murders documented in the town's first three years, the killing of gunfighter, gambler, and lawman Wild Bill Hickok is the best known. On August 2, 1876, Hickok was playing poker in "Saloon No. 10," his back to the door, when the man he had left broke the previous night walked up to him and shot him through the head. The bullet exited his right cheek, much to Bill's surprise, and then, after a few seconds, he fell forward onto the table. His poker hand left his chest and spilled for all to see—two aces, two eights, all black, and another card—a half-decent hand now known as "dead man's hand."

I buy Ads another whiskey in Saloon No. 10 and stare hazily at the chair that Wild Bill was sitting on that night. It's nailed to the wall above the front door, at home among an amazing array of other Deadwood memorabilia. There's a definite party atmosphere in the bar. The whiskey; "Mark," the guy next to me drinking Buds; and the noisy reenactment of Wild Bill's murder going on behind—all attempt to involve me, but I'm preoccupied by a problem. It's clear that culture has a system of inheritance, but there are some obvious and important differences between cultural and biological inheritance that could potentially derail the entire comparison.

For starters, when it comes to cultural inheritance, there appear to be no restrictions on who can be a "parent" and who can be an "offspring." If I grow a tremendous moustache with curls like a World War I general, and my dad copies me, *I* have become *his* parent in a cultural sense; he has inherited my moustache. This is a very different modus operandi from biological inheritance.

Then there's the question of what exactly is being copied. Thanks to some breathtaking science in the twentieth century, we now know what goes on when biological traits are inherited, to the point at which we can prove to the satisfaction of a court of law the parenthood of any one human being. We understand genes. But what is passing from one person to another when thoughts are inherited? Nothing physical seems to be handed on.

Finally, there's another difference that keeps misting up my goggles as I sit at the long bar listening to Ads try to convince Mark that English football is a better game than American football. It's a technical difference, but it is important. It carries the name of one of those people who was thinking about evolution long before Darwin: a Frenchman. Jean-Baptiste Pierre Antoine de Monet, Chevalier de la Marck, or "Lamarck" to his friends, fought in the Seven Years' War against Prussia, coined the term *invertebrate*, and believed in the Force. In fact he believed in two Forces.

Force number one was a grand master responsible for making Life more and more complex over time: a force of progress. Lamarck called this *la force qui tend sans cesse à composer l'organisation*: "the force that incessantly tends to make order." This force, he was quick to point out, was energized by the underlying laws of the universe, not the miraculous hand of God.

Force number two was a microengineer tinkering each day on billions of individual organisms—the adaptive force, a force of design. It brought about adaptation in the most logical way you could imagine: by improving those body parts that were used more and denuding those that were used less. Take the ostrich, Lamarck might have said. In the past it was not as it is now; it was a bird like any other, but, surviving on the treeless, open semidesert, it did something that other birds did not: it chose to run instead of fly. Picture those running ancestral ostriches, day in, day out, filling their silly, insufficient bird legs with oxygenated blood, stretching the muscle tissue, ripping it slightly, repairing it, every day, until, as with an athlete, each leg became laden with bulky, prime flesh. Sitting above their pounding hind limbs, their unused wings lay flat against their flanks, flaccid and fatty. Now, Lamarck would declare, the anatomical ground gained by those pioneering ostriches is inherited by their offspring. Every ostrich egg is imbued with the fruits of its parents' labor. Each new hatchling will gain a stronger leg and a weaker wing as a result of the run-rich, flight-poor lives of their mums and dads, so that over many generations, the squat, flapping ancestral bird becomes a leggy, flightless descendent one. The adaptive force has done the job of a skilled designer: sought out the parts that most benefit the immediate environmental circumstances and maximized their efficiency. What a great way to evolve!

Unfortunately, it's not the way Life evolves. It sounds pretty good when you first hear it—it explains both natural variation and inheritance—but Lamarckian evolution isn't how the biological world works. You don't inherit the characteristics your parents *acquired* during their lifetimes. No matter how often my dad played (English) football before I was conceived, it wouldn't have affected the eventual size of my adult leg muscles. It took almost a century before Friedrich Leopold August Weismann, who was German, explained exactly why. Living things have two types of cells: sex cells (sperm and eggs) and body cells (everything else). These two types of cells have two totally different functions in Life. The job of the sex cells is to pass instructions on to the next generation; they are the only parts of us that can. The job of the body cells is to ensure that the sex cells get to fulfill their ambition, by enabling their bearer to last long enough, and become desirable enough, to breed.

Weismann debunked Lamarckism by understanding that the sex cells were just cargo in a ship. Holed up in their own little rooms, safe and protected from the nasty activity of living, from the functioning organism all around them, they had no contact with the body cells. So, no matter what kind of a life the rest of the body experiences, the body cells have no way of passing that experience back to the sex cells. They cannot inform the sex cells about Life in the outside world. There is an information barrier between them.

Hence, when you pass on instructions to your offspring, the life that you have led is not represented in that act of inheritance. Your life doesn't matter at all, beyond the fact that it enabled the passing on of your instructions, that it enabled Life to gain a touch more immortality.

Depressing maybe, but also devastating for Lamarck and his adaptive force. Weismann's barrier meant that the inheritance of acquired characteristics was impossible, and Weismann's work at the turn of the twentieth century ended support for Lamarck's "soft inheritance," an inheritance in which lifetimes mattered. In its place, biological science set about building a conceptual framework around "hard inheritance," and the only theory of evolution that suited such an approach was the theory of evolution by natural selection, the theory of Charles Darwin.

I order a nine-dollar scotch. I know that I'm in the Wild West, but I can't take another bourbon. The barmaid places the bottle next to us on the bar. It's a Speyside whisky, from a distillery built before Deadwood. The blurb on the label claims that the single malt therein was made using techniques that had been handed down and honed over the centuries by the master craftsmen at the distillery. Part of my nine dollars pays for that heritage, and I'm happy with that, because it's made a big difference: the hands of those talented dead men are evident in the quality of my dram. But my problem is that the proposed Life of the noosphere now tastes a bit too Lamarckian for my liking. If I understand "honing" correctly, it implies that, over the years, the whisky distillers discovered ways of improving the drink, incorporated those techniques into their distillation method, and then passed on those improved techniques to their apprentices. But if that's so, the heritage of my whisky is riddled with the forbidden acquired inheritance, with Lamarckism. If thoughts are worked upon, improved *within* a human lifetime, and then handed on in their improved, worked-upon form, then that's Lamarckian evolution, isn't it? And isn't this how all culture evolves: people mixing thoughts together, improving thoughts, and then passing on to their cultural offspring not carbon copies of the thoughts *they* inherited but the worked-upon, improved thoughts they now possess? Isn't all cultural inheritance "soft"? Soft as a baby's bottom.

If I understand evolution correctly, this cannot be. In a Darwinian world, you can't inherit traits that way; it's against the rules. You're not allowed to pick up the hand a dead man was dealt and use it as your own. The game would fall apart. And so would Life. You have to get your cards straight from the deck. That's the rules. Is this a crack in the lenses of our new goggles?

Ads taps me on the shoulder, and I fall off my stool.

Tepee or Not Tepee

My head hurts. I crawl out of the tent, and the Harley-Davidson couple on the next pitch, with their Harley-Davidson fanny packs, Harley-Davidson towels, and Harley-Davidson T-shirts, give me a disapproving look. Perhaps I was snoring. We muster some energy, pack the tent, and head off to Rapid City. The best place to see an authentic Sioux tepee is not, I've been told, on a Sioux reservation, but in the Journey Museum. The museum is a modern monolith of brushed concrete with a tepee-shaped entrance. The day is hotter than any so far, so we fall eagerly into the shade of the museum's foyer. We skip the orientation video and head for the exhibits. There, center stage, is a proper tepee; our first one. I get my notebook out.

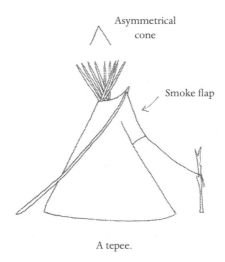

A tepee.

This much I already know: for a conical tent to be classified as a tepee it must show two defining characteristics:

1. It must be asymmetrical. The pitch of a tepee's cone is steeper at the back and shallower at the front, where the doorway is. This enables even the tallest Plains Indian to stand up comfortably at the rear of the tent and, since all tepees face east, leans the structure into the prevailing winds (racing down from the Rockies) to make it more stable.

2. It must have smoke flaps. The gathering of tepee poles at the "nest" above the living area creates a natural exit for smoke from the central fire, but a conical tent with a hole at its apex is not always practical—hence the addition of skins next to the smoke hole that can be adjusted to change the hole's shape and size. When the weather is fine and the wind is low, the flaps lie back on the flank of the tepee, decorating it. When it rains or snows, the tepee dwellers close the smoke hole by folding the flaps over each other with the aid of special smoke flap poles that bend around the outside of the tent and rest at its rear. Should the prevailing wind switch direction, the Indians lift up one of the smoke flaps so that the smoke exiting the smoke hole is sheltered and can leave the dwelling without hindrance, disappearing gaily across the plains.

All the Plains Indians had tents with these two features. My aim is to work out if tepees varied beyond that, and how and why exactly.

The Journey Museum Sioux tepee has sixteen poles crisscrossed at the nest, but by peering inside, I find that the entire structure is supported by just three: a tripod upon which all the other poles are laid. If the footprint of the tepee were a clock face, and the doorway were at six o'clock, then the tripod poles would lie at seven, immediately to the left of the door, at about ten, and at two. Above the door, six pins join the two sides of the cover together. This cover is made from buffalo hides, about sixteen skins in all. At the rear of the tent, one of the hides still has its tail on, clearly by intention, since it falls down exactly at the center, halfway up the cone. The smoke flaps are large, with pockets made from additional straps of hide at their top corners, into which the smoke flap poles are inserted.

Melvin, a Sioux museum guide, comes over as I scribble.

"Nice tepee," I say.

"Yep, cost about thirty thousand," Melvin says.

"So it's not an original?" I'm a little disappointed, but realize I shouldn't be as Melvin starts to giggle.

"No, they wouldn't last that long."

"It looks authentic, though," I pitch in.

"Yes, the guy who made it knew what he was doing; it's typical Sioux."

I'm pleased with myself: "typical Sioux" implies that other tepees are not, which implies that the different island tribes had different species of tepee.

"What makes it typical Sioux?" I press, so Melvin begins:

"It's a three-pole tepee. It's got smoke flap pockets." A-ha, two of the features I noted. "The Cheyenne tepee had the same . . ." Oh. "But they had slimmer smoke flaps, with extensions added on the bottom edge . . ." Aha! No matter how small the difference, if there was a difference between the tepees of two different tribes, it counts. Melvin goes on to show me how the skins are put together. Apart from the first skin, the one with the waggley tail, the skins are placed with the heaviest hides at the bottom, and get lighter and lighter toward the top. I'm feeling pleased as punch.

"So, who made it? An elder?"

"No," says Melvin, "a white guy, actually."

"Oh," I say out loud.

Big Country, Big Picture

Ads is snoring off the booze. The iPod is playing slow three-star tracks so that I don't have to listen too intently. We're heading back around the northern shore of the Black Hills and on to Wyoming. The midday sun, the whisky, and the worry that there are no tribespeople left who actually know how to build authentic tepees bear down on me. But mostly I'm worrying about inheritance. I have no doubt that culture is inherited—it is by definition something that is passed from one person to another—but if cultural inheritance doesn't conform to Darwinian rules, if it's Lamarckian, then a direct analogy between the biosphere and the noosphere will not wash. It would be like comparing poker with bridge; it's not the same game.

I console myself by remembering that in the field of evolutionary science, being confused about inheritance is par for the course. Throughout the nineteenth century, academics wore a furrowed brow when examining the patterns of inheritance in the living world. This was partly down to their erroneous faith in Lamarck's inheritance of acquired characteristics, and partly because they made a faulty assumption about the way in which characteristics materialize in the offspring: they believed in "blending inheritance."

The theory of blending inheritance proposed that when the seeds of mother and father met, the instructions they contained were blended such that the offspring became living averages of their parents. If a tall dad and a short mum had a baby, the baby would grow up to be medium-size. If a black woman and a white man had a baby, the baby would be brown. This was the logic, and in the nineteenth century, everyone assumed, from the evidence around them, that it was true. This "truth" worried them intently. They knew that a blending inheritance was unsustainable since it would bring an end to all natural variation. Here's how: imagine a palette of colors. Mix two together, and you get a blending of the two parent colors. Mix these mixtures together, and you get averages of them. Within only a few paint generations, every spot on the palette will be a mucky earth color; the former variation of the palette will be no more. Lamarck's theory and all the other theories of inheritance by acquired characteristics were attempts to explain how, in a world in which all heredity blended, natural variation continued (which it clearly did; ask any butterfly collector).

Darwin was one of those wearing a furrowed brow. He dedicated years to trying to resolve the general confusion. His best attempt was a theory named pangenesis, which he published in 1868 in *The Variation of Animals and Plants Under Domestication*. In essence, pangenesis was no more than Lamarckism 2.0. The only thing Darwin added to Lamarck's "theory of use and disuse" was a stab at the physiology that might lie behind it. He proposed that each body cell produced "gemmules," heredity particles that were delivered to the genitals, loaded within sperm and eggs, and subsequently activated in the offspring as they developed. This *updating* of sperm and eggs as the parents continued their lives would, Darwin believed, help to account for the influence of the environment upon the adaptation of organisms.

He must have been desperate. Siding with Lamarck on this subject undermined his own case for natural selection. It implied that the responsibility for bringing about adaptation was shared by the two mechanisms. But his desire to solve the inheritance mystery overtook him. Even when pangenesis was disproved—by his half-cousin Francis Galton, who, with the best will in the world, could not find gemmules in the blood of rabbits—Darwin wouldn't admit defeat. He suggested that the gemmules must be carried by another, yet undiscovered transport medium. In the end, inheritance would confuse and worry him to his grave.

The tragedy is that, as early as 1866, only seven years after the publication of *On the Origin of Species*, the truth was already out there. Written in a language few people could understand, published in a journal few people would read, a paper by a Silesian monk called Gregor Mendel gave the first tantalizing details of the science of the "gene." By meticulously crossing pea plants and noting which traits were inherited by which offspring, Mendel discerned the mathematics of biological inheritance. The figures suggested something that would have immediately brought the nineteenth-century inheritance debate to an uneasy pause: *there is no blending in heredity.* Mendel's peas told him what we now know to be true: that we inherit from our parents particles of heredity carrying instructions that will bring about traits; that the offspring of sexually reproducing species get two copies of each particle, one from their father, one from their mother; that there is a competition between these particles to be expressed; that dominant particles win, and recessive particles lose; but that although recessive particles lose, they don't disappear, and will themselves be carried on to further generations in further sperm and eggs, with the possibility of winning their right to be expressed in the future.

All this insight lay fallow until after both Mendel and Darwin's deaths. When it was rediscovered and announced in 1900, Mendel's inheritance was found to blend beautifully with Darwin's natural selection—the particles were even renamed "genes" in honor of the "pangenes" of Darwin's discarded theory—and the science of biology was propelled into a century of incomparable illumination. At the halfway mark, in 1953, Francis Crick and James Watson walked into The Eagle pub in Cambridge, England, on a cold winter's day and yelled, "We've found the secret of Life," and they had. They'd discovered the structure of DNA, the molecule that hosts the particles, and as a consequence, they found out what a gene actually looked like. Fifty years after that, in 2000, the Human Genome Project published a "rough draft" of the entire human DNA sequence, the instruction manual for humans, what *every* human gene looks like. One hundred and fifty years after *On the Origin of Species*, we're a lot less worried about inheritance.

But this doesn't mean that we get any more sleep. Why? Because, in discovering the truth behind biological inheritance, we have come to realize just how complex it is. Why do black women and white men have brown babies? We don't know exactly. But we do know that the tone of human skin is determined by seven or more different genes, all of which appear in a number of different forms (or alleles), each of which is variously dominant or variously recessive, and all of which interact in complex ways to trigger the production of melatonin, the black pigment in skin cells. We may now know that the inheritance of skin color is not blended, but that doesn't make it any easier to explain. We'll have to work nights.

And the inheritance of most traits in the biological world is just as involved. I don't even come close to having my mum's nose. I have a natural variant of both my mum's and dad's noses, which came together upon the hard work, interaction, and negotiated expression of dozens of their genes, nose building within the constraints of what nose-building materials were available at the time and impeded by the damage I was doing to my nose as it developed (falling upon it while learning to crawl, etc.), with the possible additional influences of random genetic mutations that may have (accidentally) added to the mix. My nose was raised like a barn was raised. The only reason it looks more like my mum's than my dad's is because her instructions, to some unknowable degree, were being shouted louder than his during the nose raising.

Inheritance is a complicated thing. Life is complicated! Staring at noses won't reveal its secrets. And staring at whiskey bottles in bars won't either. I've got to be patient, resist making assumptions. The truth may already be out there.

"Welcome to Wyoming!" Ads shouts. I whack on the Aaron Copland.

6

Selection, Wyoming

Mindless[+]

Wyoming is a soft gold color. The hills roll again, sporting the familiar crew cut of the shortgrass, but they are now also speckled with sagebrush. Coal trains glide by on the parallel railroad. They are so long and slow that they assume a degree of permanence in the landscape. Ads and I get used to their being there. After some time we realize that we never see their engines, and we never see their ends; the trains exist only as chains of wagons, appearing and disappearing courtesy of the coulees. It makes Wyoming look like a glorious stage set with layers and layers of gently animated scenery.

The drive, as a result, is good for our hearts. We gladly stick in cruise mode, as do the other drivers on the road, and the almost imperceptible shuffle in traffic enables me to study the Wyomingites. The baseball caps are gone. Everyone here has cowboy hats, and I suspect it's going to be cowboy hats for some time. Just as grubby farmers appear to be positively associated with pronghorns, cowboy hats are positively associated with cattle ranching, and cattle ranching is clearly the dominant occupation of the farmers of Wyoming. But why should that be? Where did the cowboy hat come from to assume such dominance in this part of the world? Why is it that all the folk from 'round these parts (and not the parts a couple of hours east) should *select* to wear cowboy hats?

Selection is the third and final essential ingredient of Darwinian evolution. Once natural variation is present and a system of inheritance is in place, it is selection that implements the design itself. This is the bit that Darwin *did* come up with, but, in hindsight, he could have selected a better term. To many, the word *selection* implies that there must be some kind of mind at work behind the action, since selections are, by definition, intentional, conscious acts. That, of course, was not Darwin's idea. *Selection* also suggests that nature is picking "the best," when in fact it does this only by default.

Darwin originally chose his troublesome term in order to highlight the similarities between what nature does to wild organisms and what humankind does to tame ones. Let me give you an example: 'Round these parts, cattle breeding is a serious business. Semen from a bull with the right pedigree can fetch many hundreds of dollars a straw. That's because the DNA of that bull is chockablock with only the best information. His calves will be delivered without a hitch, grow fast, and taste good, and because of that, the farmer's profits will do the same. The farmer selects the variety of bull that has the traits he would most want his cows' offspring to inherit.

To Darwin, this activity of "artificial selection" was an exact parallel of the work that nature was engaged in out there in the wilds. Indeed, the only significant difference is the speed at which artificial selection can work. Because the mind involved has the luxury of picking the best, rather than picking *out* the worst, the evolution of traits over time occurs at a far greater rate. Cattle, pigs, dogs, chickens, bananas, potatoes, wheat—all the organisms that we have involved in our special form of survival of the fittest have thus experienced hyperspeed evolution; they've changed enormously over a relatively short period of time. But it is still Darwinian evolution; the selection still brings "survival of the fittest."

So the "mindless" tag that I've often added to Darwin's evolution is, I have to admit, a bit misleading. Darwinian evolution *can* bring about design mindlessly—that is its unique skill, and indeed, before we showed up, it was its only option—but the involvement of mindful decision making does not preclude Darwinian evolution. If a "choosing mind" is selecting from a range of varieties with inheritable traits, then all it's really doing is taking the place of nature in the equation, rather than coming up with a whole different form of mathematics. An alien surveying the evolution of the Earth's biosphere would have to conclude that it was 100 percent Darwinian, but only 99.9 percent mindless. Perhaps "mindless[+]."

However, that's not to suggest that every bit of selection we are involved in is intentional or conscious—far from it. Take my dog, Molly. The evolutionary journey of her species over the past twenty thousand years has been intimately related to the actions of our species, but until recently it wasn't intentional.

Dogs were the very first domesticated organisms. They diverged from wolves tens of thousands of years ago, probably in the Middle East. The consensus is that they began their program of domestication accidentally/automatically when, as wolves, they started to hang around hunter-gatherer tribes. The dogs with the shortest "flight distance"—those least likely to run when humans appeared—were able to scavenge more food. Hence natural selection favored a trait that we might call "tameness."[1] Coupled with this tameness trait were other characteristics: small size was selected, the wolves' coats changed color, their jaws shortened, their teeth shrank, and they started to act like puppies even when adult— wagging tails and barking—and they actively sought human touch and returned eye contact. In short, they became dogs. All of these traits were positively reinforced by their human companions: the more doglike an individual wolf, the more they were fed by the tribe members, and the more they survived and reproduced—but this was still mindless self-design.

These early dogs, it has to be said, looked nothing like my Molly. She's a cocker spaniel, and her features are the result of the centuries of the less accidental/more intentional "selective breeding" that took place subsequently. Cocker spaniels are "flushing dogs," designed to drive ground-dwelling birds into the air so that hunters can blast them out of the sky with guns. Molly has exactly the features I would need in a dog should I wish to hunt woodcock. She is low to the ground; she loves diving through undergrowth and getting into water. She's ever keen to please. She uses her nose permanently to find interesting things. She zig-zags as she walks (given the chance), and she likes nothing better than to retrieve the carcasses of fallen birds.

These features didn't come together by accident. They were brought about by the discerning work of generations of English dog breeders. But don't be under the impression that the breeders were intentionally / consciously breeding a "cocker spaniel." All they were doing was intentionally / consciously permitting those dogs that proved to be the most effective at flushing woodcock the right to breed, and denying those dogs that tended to frighten the birds before the guns were ready, or that were less keen to retrieve the quarry, or that were scared of water, the same privilege. The result was a dog like Molly—but a dog like Molly wasn't their intention exactly.

However, now there *are* breeders in operation with exactly that intention. They want a Molly. More, they want the *ultimate* Molly. The craft of dog breeding has moved on again. Each year, in March, the ultimate in cocker spaniels are waltzed down the catwalk for judges to scrutinize at the United Kingdom's biggest dog show, Crufts. If they don't display the idealized cocker spaniel features, they lose. But if they demonstrate a "square muzzle, with distinct stop set midway between tip of nose and occiput," "full, but not prominent eyes," a body "neither too wide nor too narrow in front," and dozens of other decreed traits, they have a chance of winning the best of breed, and going on to bear scores of uber-cocker spaniel puppies themselves. How will this end? Well, there are already two forms of cocker spaniel recognized: the "working cock-ers," which still look and behave a little like their flushing ancestors, and the "English cocker," which is primarily a show dog, with all the right proportions and a temperament conducive to being popped up on a po-dium to have its bits measured.

So the relationship that Molly's kind has had with our kind is marked by at least three different brands of artificial selection. At the outset, twenty thousand years ago, our participation was accidental / automatic and thoroughly mindless, just as normal natural selection is. During the time at which her breed was being crafted, there was a degree of intention, but the dogs themselves were still allowed a modicum of self-design. More recently, with the advent of pedigree breeding, the fate of traits has become entirely human controlled, and ultimately mindful (not that, to my mind, there is any real point to it).

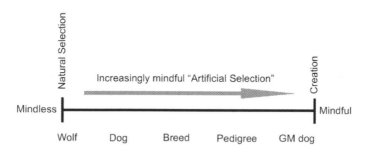

The mindlessness-to-mindfulness spectrum.

I want to place these three brands on a new imaginary device of mine, the mindlessness-to-mindfulness spectrum. This spectrum is a single line along which lie all the conceivable parties empowered with the capability of generating living things. All that changes as you move from one end of the spectrum to the other is the degree to which the party is mindful. At the extremes of this spectrum are, on the left, the most mindless thing imaginable: Darwin's pure, old-fashioned natural selection. On the right, the most mindful thing imaginable: the Judeo-Christian God.* In between these two extremes lies the sliding, varyingly mindful world that we tend to bracket under the title "artificial selection."

Molly the cocker spaniel's twenty-thousand-year evolution entailed a dramatic slide from left to right along this spectrum over time. Beginning at the extreme left, her ancestors were originally within the jurisdiction of pure natural selection, a time in which they were known as wolves. But as soon as their growing relationship with human beings had an effect on their fitness, the goal posts changed, and Molly's ancestors stepped on to the sliding scale. As they slid to the right (to dogs), through selective breeding (breeds) and on to pedigree breeding (pedigrees), they would have evolved at a faster and faster rate. That is the consequence

*Other Creators are available.

of an increasingly mindful involvement. Will her kin ever get to the extreme right-hand "godlike" position?

The extreme right-hand end is a special place. It's the only point on the spectrum at which Darwinian evolution—and, indeed, evolution of any kind—has no rule. It is the point of "creation," where the ultimate in intentional/conscious minds has the power to instantaneously generate Life. It is true that our work with organisms is getting closer to that right-hand end; we are beginning to "play God." Genetic modification is often described in those terms. Where does GM fit on my spectrum? Just shy of the right-hand end. It's a sort of "artificial creation." It's not instantaneous, and as its title suggests, there is a prerequisite for existing material. Even the most adventurous GM practitioner is confined by the genes offered up by natural variation, but that's all they are confined by. Inheritance is no longer a problem; that can be manufactured. In fact, that's the difference between artificial selection and artificial creation. In the former, while selection is artificial, inheritance is still natural. In the latter, the inheritance has also become controlled by minds. Presumably, all we have to do now is artificially manufacture the variation—build never-before-seen genes*—and who knows, we may achieve creation proper. But even then, the "creation" itself, the living product of all that work, will still be subject to Darwin's evolution. In Life, there's no escaping it!

The Evolution of the Cowboy Hat, Served Three Ways

If our species' varyingly mindful approach to domestication has recently made the evolution of one corner of the biosphere so complicated, how much more complicated must it make the evolution of the noosphere. After all, the world of human thought has no other option but to be intimately involved with our minds. Well, perhaps it's not quite as intimately involved as we would like to think. Just because thoughts travel through our minds on their journeys to others, it doesn't necessarily mean that we are *mindful* of them.

You say po-*tay*-to, I say po-*tah*-to. En masse, pronunciation differences are referred to as accents. Ads and I have no accent when in England,

---

*Recent "synthetic biology" work by Craig Ventor is already well on the way.

but here we are heavily accented. We've already received a few quizzical looks when asking questions of locals on this trip. Some people pick up the clues in our words and joyfully pronounce, "You're from England!" but many more can't place us. Once, when I was in Seattle, a woman asked, "Exactly what state are you from?" I told her I was from the East. I was once in Alabama when a dad rallied his many children with the cry "Hey y'all come 'n listen to this guy. He sounds like an alien or somethin'," and then made me recite nursery rhymes while they variously laughed or cried, depending on their age. We don't mindfully choose our accent, dialect, or language. We inherit them mindlessly from the people who surround us when we are young. On the mindlessness-to-mindfulness spectrum, the selection of accents, dialects, and languages is at the extreme left, akin to pure natural selection in the biosphere. Over time these evolve all on their own, changing significantly without any intentional/conscious participation on our part.

What else is like that? Back to the question of the cowboy hat and how it came to be so dominant in Wyoming. There are three answers to that question, and the answer you give depends on where you site the evolution of the cowboy hat on the mindlessness-to-mindfulness spectrum. And this, in turn, depends on how much you squint and whether you're wearing goggles . . .

ANSWER ONE: STETSON INVENTED THE COWBOY HAT

Every cowboy will tell you that John Batterson Stetson invented the cowboy hat while digging in the dirt in Colorado at the start of the 1860s. Chased west by tuberculosis as a young man from his native and overcrowded New Jersey, Stetson ended up in the equally squalid shanties of Pikes Peak during the gold rush. As a son of a hatter, he soon noticed that the unique lifestyle of the West was short of bespoke headwear. The prospectors around him wore a pitiful assemblage of hats leftover from previous incarnations, and none of them was perfect. So, in between digs, Stetson began experimenting with the furs available in the gold rush towns. He used the techniques that his dad had taught him to bash them into a fur felt that was waterproof. One day, cursing the sun on his aching back, he stitched together a whole batch of these fur felts and made an overly wide-brimmed hat with an overly tall crown. He decided to wear it even though it looked ridiculous; fashion meant little on Pikes Peak. It was stiff, waterproof, and cool (because the high crown let John's hot head breathe). He returned east without gold, but with a good plan.

Original Boss of the Plains.   Modern Stetson.

In 1865 he began to manufacture a hat that resembled the one he made that day. It was wide-brimmed, made of fur felt, with a crown of four inches. He called it the Boss of the Plains. It was the first hat manufactured especially for the cattlemen of the Wild West, and it was the dullest-looking cowboy hat you could possibly imagine. The crown was even and rounded (no peaks or dents), the brim was flat (no rolled edges), and it bore the drab gray color of the fur it was made from. However, as the buying public couldn't know what cowboy hats were to become, the Boss of the Plains took the West by storm.

Stetson's hat company rapidly grew into one of the largest in the world. He never ceased the development of his designs, inventing many other classics such as the Carlsbad and the "ten-gallon hat" that was eventually to appear on the silver screen, sealing the image of the traditional cowboy forever. By the time Stetson died, his company had more than five thousand employees and made more than three million hats a year.

(On the mindlessness-to-mindfulness spectrum, answer one, the textbook answer, lies well over to the right. Stetson was an inventor, and his invention of the cowboy hat was an act of artificial creation, as with the genetically modified cocker spaniel, drawing on existing materials, but making an almost instantaneous giant leap nonetheless.

Time to slide left . . .)

ANSWER TWO: COWBOYS INVENTED THE COWBOY HAT

In fact, what happened, of course, was that no one person invented the cowboy hat. Stetson was important because he was the first to mass-produce a wide-brimmed, high-crowned hat, but for many years before he helped to dig up Pikes Peak there were hats in circulation that looked very similar to the Boss of the Plains. What secured the cowboy hat as an icon of the West was not Stetson but the homemade ingenuity and subsequent purchasing preferences of the cowboy clientele. It's simple capitalism.

Drawn from the East Coast and beyond to the Wild West by the promise of gold, furs, and land on which they could raise cattle, a population of frontierspeople found themselves in the middle of the untamed continent, in the middle of the century. The sun, rain, wind, cold, and aridity would not leave them in peace. They sought sanctuary in any hat they could hide under, but their previous lives had left them ill-equipped. Walking around any western town at that time, you would have seen an odd assortment of hats sitting uncomfortably upon hapless heads, none of them right for the West. The silk top hats that were the height of fashion back east were entirely unsuited to a hot, dirty, horse-riding life. Homemade coonskin hats kept your head warm, but they filled with fleas, and the sun burned out your eyes underneath. Derbies protected your head while you were on a horse, but not your face, which grew lobster pink. By far the most suitable hats were the wide-brimmed straw "planter's" hats, felt "slouch hats," and the Mexican sombreros of workers who had migrated in from other lands. They kept the sun out of your eyes, but they were poor in the rain and hot to wear.

The flowering population naturally adapted its headgear. Worthless hats were cast away, and wide-brimmed hats became standard wear. They were, after all, the natural selection. Hides were readily available, and many people did what they could to fashion waterproof hats from leather and fur. They were functional but inelegant, insufficient for the increasingly civilized West.

In the early 1860s, the Civil War raged to the east. Spilling from its skirmishes, the "Hardee hats" of Union soldiers and "stag hats" of Confederate cavalry—both wide-brimmed, with high crowns—found their way into the general population on the heads of former soldiers. Stripped of their badges and feathers, the hats made good riding cover in the hot country, but did not improve the millinery of the frontier. This was still a land of bastardized homemade and secondhand headwear.

Then, from 1865, appearing in the stores on Main Street, a unique addition: the Boss of the Plains, a hat made back east by one J. B. Stetson. It fitted the Wild West like a glove. It was shaped like some of the existing western creations, but it was much better made, crafted from quality fur felt, which ensured that it was entirely waterproof; and most important, it was new. The townfolk clamored for it. An elegant, civilized western hat—with a great name! Buyers wanted to be "bosses of the plains." Compared with the other hats in circulation, the Boss was pricey—at least half a month's wage to a cattle driver—but because it was pricey, it became a status symbol; and because it became a status symbol, it was extremely desirable. Within only a few months every frontiersman and cattle driver had to have one.

But also, because of the price, *one* was all each cowboy could ever have, so they used their Stetsons well, and in ways that Stetson himself would not have imagined:

My Old Stetson Hat
(author unknown)

Stained with alkali, sand and mud
Smeared by grease and crimson blood.
Battered and bent from constant use,
Still you stood up to all the abuse.

A true companion through all these years.
Fanning broncs and longhorn steers.
I dedicate this to that ol' grey lid,
For the useful things the old hat did.

Used to decoy some rustler's lead,
Or as a pillow 'neath my head;
Coaxing a smouldering fire in the cold,
Panning dust in search of gold.

Pushed up big, then knocked down flat,
Has been the lot of my Stetson hat.
For carrying oats to a piebald bronc,
Security for drinks in the honkey tonk.

Mistreated, abused on a roundup spree,
Walked on, tromped on, old J.B.
Fighting fire in a clapboard shack,
Or stopping wind through an open crack.

Been everywhere a hat can go,
In 48 states and Mexico.
I've grown old as we trailed along,
While you, old hat, are going strong.

You've been a good pal through all of that,
You dirty, old, grey Stetson hat.

For all these reasons, Stetsons, with continued use, came to assume a new physique (dare I say, just like Lamarck's ostriches). Their crowns became more and more dented with each "yee-ha!" Their brims became more rolled at the flanks with each night under the stars. On the long cattle drives up from Texas to the railheads where the cowboys would load their meat onto trains bound for the eastern cities, the smooth, elegant Boss of the Plains hat would degenerate into something that was creased, dented, rolled at the edges—something, in fact, that looked much more like a "cowboy hat."

In those railhead towns, envious of Stetson's success, rival firms set up to manufacture western hats. The pressure to get the product exactly right intensified. The growing population of southern cowboys began to have their say. Primed with a history that had been smattered with Mexican sombreros, they eagerly bought any fur felt hat that arrived with a bigger crown and a wider brim; that sun was hot! And to contrast the choice of townfolk who had never known the true life of the plains, and who still had their Boss of the Plains neat, smooth, and elegant, the cowboys chose any hat that appeared with decorative dents in the crown and a prerolled brim, because those were the hats that most represented their lives, those were the hats on the heads of their homegrown heroes.

This persistent and discerning hat selection by the cowboy clientele, guided by a still-evolving western aesthetic, crafted the true cowboy hat, the one we all recognize today, the one that Hollywood told us existed years before it actually did.

(Answer number two slips and slides about the center left of my mindlessness-to-mindfulness spectrum. In this context, Stetson's "invention" was less an act of creation than an adaptive move of above-average mindfulness, a timely tweak of existing models. His hat was not a new species but a new morph within a widely variable population of headgear that was already subject to the selections of the diverse western public. Ultimately, it was their varyingly mindful selections that brought about the evolution of the cowboy hat.

Left again? How can that be?)

ANSWER THREE: THE COWBOY HAT INVENTED ITSELF

There is a third way to answer the question "Who invented the cowboy hat?" After sailing through these four states, feeling small in this vast space, and seeing how the noosphere can boast natural variation and seems to have a system of inheritance in much the same way as the biosphere, it's an approach of which I'm becoming more tolerant. In essence, it's similar to answer number two, only the focus is switched. Instead of focusing on the community of cowboys who, together, selected the cowboy hat, it depersonalizes the story and proposes that the cowboys were *just a front*, a collective of minds that formed a unique selective environment in which populations of free living thoughts existed and evolved and, in the process, as a side effect, brought about the self-design of the cowboy hat. You may, quite possibly, hate answer number three, because it casts an act of cultural evolution at the forbidden end of the mindlessness-to-mindfulness spectrum, but I'm just telling you what I now see through my new-world-view goggles. Here it is, using terms normally reserved to describe the evolution of Life in the biosphere:

When people first moved to farm the Wild West, their heads were hosts to various species of hat design. Some of these were manifest on their heads, doing a terrible job of keeping them protected from the sun, rain, wind, cold, and aridity, but there were other designs, invisible ones, hidden within their collective repository of thoughts: the tricorn perhaps, the beret, the deerstalker, and so on. Very few of these designs were fit for the environment here.

Within a generation, the pioneers defaulted to wearing only a tiny subsection of the "hats" in their heads on their heads. These were hat designs with traits that best suited the prevailing environment—the wide-brimmed planter's hat, the slouch hat, and the sombrero—but none was perfect. They weren't waterproof. The planter's hat and slouch hats were hot to wear. The sombrero was too wide-brimmed: its design was adapted for the harder Mexican sun; up north, the unnecessarily large brim just got in the way. Natural selection worked on the three hat designs. In each of the three "hat populations," the hats born with more suitable traits were copied more often than the hats born with less suitable traits. As each subsequent generation of hats came into being, there was a gradual adaptation in form toward an ideal "western hat" design. In evolutionary biology, this type of selection is termed "directional selection." It applies to any bout of adaptation in which the entire population of a species gradually moves in the same direction over time. Of course, the three hat designs were coming from different directions, but since they all had the same ultimate destination, the new niche of "a hat fit for a cowboy," their designs began to converge upon this one ideal.

However, before they could achieve it, "out of nowhere" a new species arrived that made manifest that ideal form: the Boss of the Plains. In fact, the Boss of the Plains wasn't the ultimate cowboy hat—it, too, would come to adapt—but at that point in history, it didn't have to be; it simply had to be fitter for the selective environment than its three competitors. It was, and as a result, its population flourished.

And as its population flourished, a peculiar "density-dependent effect" came into play that gave a further boost to its success. The selective environment was such that as the density of the new species increased, its relative fitness also increased. In regular-speak, we'd call this density-dependent effect *fashion*. Upon the successful establishment of the vanguard (which Stetson cleverly guaranteed by sending free batches to all the big retailers), and exploiting this fashion effect, the success of the Boss of the Plains in the new niche was inevitable. In only a few years it all but wiped out the competition.

But, with time, the niche itself grew, and the growing complexity of the selective environment brought a "disruptive selection." The two different clienteles, town and country folk, exerted their influence upon the population of Boss of the Plains in slightly different ways. Where natural variation occurred—a rolled brim here, a dent there—the trait was either selected *for* or *against*, depending on the nature of the selector examining the design. The Boss of the Plains design inevitably diverged and came to exist as two morphs: one morph, which resembled the original, was more prevalent in urban areas; a second, with dents and rolled brims, became more abundant in rural areas. As it happened, this rural morph, born far and wide on the heads and *in* the heads of their host cowboys, was to expand its range significantly throughout the West.

Within a matter of years the broader selective environment of the

frontier was populated with the cowboy hat design in all its subtle varieties. But, as a result of this species "saturation," the selection pressures
increased on all traits of the design. Traits that didn't matter that much
before now made a difference. The dents had to be just right, the color
was important, the roll of the rim and the material of the sweatband
could mean success or failure for a new hat design. Each partially isolated
locale—Amarillo, Texas; Cheyenne, Wyoming; Butte, Montana; Calgary,
Alberta—presented its own unique selective environment and forced
regional adaptations accordingly. With little or no outside influences, the
emerging styles of those regional varieties were reinforced over time,
until a cowboy from Montana and a cowboy from Wyoming could be
told apart just by the lines of their hats.

This was the selection story up until the twentieth century. The advent of Westerns, country singers, and fancy dress has complicated cowboy hat evolution immeasurably since then. Who knows what a cowboy
hat will look like in fifty years' time?

Well, I know the answer to that: no one, because no one person is
designing the cowboy hat. The cowboy hat is mindlessly *self-designing*,
just like hammerheaded fruit bats and Professor Gerald Joyce's RNA
molecules. The communal nature of selection renders any minds that
are (varyingly) involved in the evolution of the cowboy hat null and void.

Stopping in Sheridan for a coffee, Ads and I spy racks of the latest
morphs hanging in the window of a western apparel store, exaggerated
production forms of the abused late-nineteenth-century Stetson: curly
rims, pursed crowns, high and wide. I wouldn't mind getting one, but I
wouldn't know where to start. I don't really know what I like. There's a
whole range of them that are okay. But inside the shop is a real cowboy
trying one hat after another. He's Goldilocks when it comes to cowboy
hats: only one of the dozens on offer will be *just right*. We watch him
through the steam that rises from our lattes as he undertakes his selection, mindfully. He takes his time and, after plenty of deliberation, purchases a hat with a high rolling brim. Why did he select that hat? Was it
a perfectly respectable, practical reason: "it must roll high at the sides to
guide the rain onto my horse rather than me"? Perhaps it was a yearning
to be like someone he admires: "it must roll high at the sides just like
rodeo star Trevor Brazile's." Or was it a from-the-gut, hard-to-define feeling that simply favored one trait over all the others: "it must roll high at
the sides"? The point is this: in the end, it doesn't matter. The reasons he

chooses a hat with a high rolling brim have no direct bearing on cowboy hat evolution. What matters is the selection he makes: a hat with a high rolling brim. The action of that selection is the only thing that makes an impression on the evolution of the cowboy hat design as it continues its almost imperceptible shuffle into an unknowable future.

Ultimately—forget the fancy scientific terms—this is the only real difference between answer two and answer three: in answer two, we are claiming that bit by bit the cowboy hat became what the cowboys *wanted* it to become; that their communal intentional/conscious decision making brought about its design; that they, ultimately, were the architects of their cowboy hat. While, in answer three, we declare that the consciousness or otherwise of their decision making is not the point; that whether cowboys chose a hat because it would "decoy some rustler's lead," or because it would act "as a pillow 'neath my head," or because it could coax "a smouldering fire in the cold," or because it was perfect for "panning dust in search of gold" is all by the by. In the end it was the *patterns* of their selections that guided the future evolution of the hat, not the *reasons* for their selection. One hundred and fifty years later, the glittery, pink cowboy hats that sit atop brides-to-be as they fall out of bars on their bachelorette nights retain the form demanded by those patterns of selection, but the reasons for those selections are lost.

The Idea

The hats that cowboys wear *could* have evolved just as mindlessly as the accents that cowboys use. We have the impression that invention, intention, and conscious decision making play a crucial part in crafting our humanity, but perhaps that just isn't so. Perhaps the evolution of the noosphere can only ever be described as mindless[+].

Four states in, and cultural evolution is beginning to look remarkably similar to biological evolution through these goggles. Culture appears to have natural variation like biology. It has a system of inheritance (a bit) like biology (and I'm still working on that). It has mindless[+] selection like biology. Through these goggles, culture *is* like biology—not exactly the same, but alive in the way that biology is alive, with a proper, capitalized Life. I can feel the Force!

And in feeling the Force while sailing across the empty plains, I've had a further revelation: I was right to suggest that we humans span two worlds, but I was wrong to imagine that we *inhabit* them both. We inhabit only one world, the world that bore us: the biosphere. The other world, the noosphere, is a world housed entirely within our minds; it inhabits us! Staring at barns and hats and whiskey bottles, I've come to appreciate that this new world is in fact inhabited by a completely different type of living thing, a superphysical living thing for a superphysical world: our thoughts. Although, since they skip between our minds and between our generations, and orchestrate their own change, they appear to enjoy a degree of independence from us. They seem to lead a Life of their own. And what do you call a thought with a Life of its own? An Idea.

So, this is the idea: in the noosphere, a wilderness analogous with the biosphere, with its own Darwinian wild Life, Ideas are the living things, the "species." They are not the barns or the hats or the lutefisk, but the thoughts behind those things. Anything, in fact, that can be memorized and handed on from one mind to another will qualify as an Idea. If it's not the design of a cowboy hat, it might be a recipe for bouillabaisse, or a contribution in a brainstorm, or a method of folding paper to make a paper airplane, a way of pronouncing the word *potato*, a juicy piece of gossip, a new tune, a hand gesture, a joke, a style of jeans, a brilliant sociopolitical philosophy, a technique for undenting a Ping-Pong ball, a plan for a funny new roof on a barn, a set of instructions on how to build a tepee, or indeed, an approach to explaining the word *idea*. In this context, they're all Ideas with a Life of their own.

But how do these living Ideas actually live their lives? I've been thinking about this, and I want to introduce my idea of the Idea with a few examples of one of the Idea types just listed: the joke.

1. Q: What's brown and sticky?
   A: A stick.
2. Q: Did you hear about the marriage of two television antennas?
   A: The wedding ceremony was a real drag, but the reception was fantastic.
3. I went to the gym the other day because I wanted to join a yoga class. The instructor asked me how flexible I was. I said I could do Tuesdays.

Three different jokes from three different places. Someone told me the first one ages ago. It amuses me, and the kids love it. I think I heard the second one at a wedding, conveniently enough. The third one, I read on the Web.

My request is that we now picture each of these three jokes as a different species of Idea. Let's think about how they exist in the noosphere. Across the world, there may be hundreds, if not thousands, of people who know joke number one. In each of these people, the joke will exist as a portion of devoted memory. The correct way to perceive this situation is to recognize that each of those jokers has a *copy* of that joke in his head. From head to head, there may be variation in the copies. I've seen a version of joke number one that asks, "What's long, brown and sticky?" This is a variant on the joke I inherited years ago. But natural variation is bound to happen in living things. It doesn't necessarily mean that we should think of the "long, brown, and sticky" joke as a new species; you wouldn't tell both at the same party, would you?

To use the terminology of biology, each copy of the joke (even with a little variation) is an "individual" of that species. The total population of that joke will be equivalent to the number of individuals in existence. This may be more than the number of heads hosting the joke. For instance, my head, and now your head, hosts two individual joke number ones: the "standard" morph and the "long, brown and sticky" morph. But regardless of how many host heads there are, the more individuals of any one Idea there are, the bigger the population is and the more successful that species of Idea.

How does an Idea increase its population? In exactly the same way as the living things of the biosphere: by maximizing both survival and reproduction. In this new world, "survival" is the amount of time an individual Idea can remain in the memory of its host. This, after all, is the "lifetime" of an individual Idea. Human memory is a remarkable thing, but it does have its foibles. For an Idea to be successful, it will need to navigate those foibles in order that it may remain in the front of the mind for as long as possible.

But living long in the host is only half the battle. No human mind sparks forever, so all individual Ideas are also mortal, and every individual is compelled to "reproduce." An Idea reproduces by exploiting our Dennettian talents and spawning "offspring," exact or near-exact copies of itself, in the heads of other humans. This is a tricky operation—it requires our participation—so the Idea must somehow evolve strategies that enable it to grab spawning opportunities when they come along, and (make us) make the best use of them.

Each of the three jokes I've recounted takes its own approach to maximizing survival and reproduction. I sympathize with them. They don't have an easy life. There are thousands upon thousands of other jokes competing to get into our memories; the stakes are high. And even if they manage to get in, most don't survive for long. I find jokes really hard to remember. I hear loads of them, and I want to tell them to other people, but I often find myself stuck in the telling, with jokes half-remembered, until my audience drifts away to do something else, anything else, clearly determined never to inherit my joke.

Why are jokes so hard to remember? I think it has something to do with the difficulty of the craft of joke telling. Jokes work only if the joker is able to elicit a reaction of surprise (pleasant or unpleasant) in the jokee; "you take them one way, and then you take them another." This means that there's a climactic moment to hit, and hitting that climactic moment requires three things: the best possible wording, the best possible expression, and the best possible . . . timing. That's a lot of information to remember, and if you can't remember it accurately, then subsequent joke-spawning events will fail, as mine often do.

The fact that the noosphere is a tough place for jokes means that all jokes will be under tremendous pressure to adapt. Ideas adapt in exactly the same way that the species of the biosphere adapt: their natural variation is tested by the selective environment, and the fittest go on to form the subsequent generation of that Idea. So having natural variation is essential. Ideas get their natural variation primarily from mistakes that occur in those fragile spawning moments when one Idea is copied from one brain to another. In the biosphere, these copying errors are fairly common. We call them mutations. Most mutations won't be catchy enough to be imitated and passed around, but some may be memorable or funny enough to merit a new lineage of jokes.

Without casting your eyes back to the three jokes, can you remember them? My bet is that you'll remember joke number one most easily because it contains the least information. You may remember most of joke number two—its Q&A format, its general content—but if you try to tell it to the nearest person now, without thinking, I bet you'll run into the sort of problems I run into. Joke number three, I would suggest, is the hardest to remember and to tell well. I've had to rehearse it in order to get it just right. It takes that much effort on my part to encourage it to survive and reproduce in my befuddled mind.

You can see how natural variation will arise in jokes number two and three. Most of your copies will be mutants, and most of those will fail to survive and reproduce as a result. It's tough, but then, as Darwin realized, the "struggle for existence" is what creates the Force in the first place. Life is no joke.

Struggling to survive, struggling to reproduce, adapting over time—this is how Ideas exist in the world created between and within our minds. They evolve like independent beings, designing themselves mindlessly, accidentally/automatically as their generations navigate a path, *any path*, through the selective environment we consciously/subconsciously/unconsciously create.

The noosphere is not "a world of our own," as I formerly believed; it's an "Ideosphere": a world buzzing with millions upon millions of living Ideas. What we refer to as "human culture" is merely the sum of their bit parts. And the Life I now think I see in the noosphere is simply an epiphenomenon of their ceaseless, frenzied melee.

Anyway, that's the Idea.

PART III

History Lessen

7

Mind Out?

Goggles Off

These goggles are starting to give me a headache. They're becoming uncomfortable to wear, too restrictive. Or should it be too "reductive"? "What we refer to as 'human culture' is merely the sum of their bit parts." That can't be right. The goggles have gone too far. I'm taking them off for a bit.

We're heading north through eastern Wyoming. We've encountered real mountains for the first time. They loom in the haze of a brilliantly lit day. I can just see forests and high meadows bathing on top. Steel gray cliffs glisten below. We've crossed the plains and reached the vanguard of the Rockies: the Bighorns. They take me by surprise. I wasn't expecting to see such uplands this far east. After all this flat land, they tantalize. Ads and I want to explore them, clear our heads, but we're not diving in just yet. Tomorrow we have a date in Montana, with the self-billed "Tepee Capital of the World." So we drift north, in parallel with the Bighorns' long flank, repeatedly glancing at them as though we're deep in conversation.

Ads is driving, which allows me to switch off for a bit and default to "ponder."

I can see that, because we are only human, much of the variation of Ideas in the noosphere has resulted from our past accidents, that at least some of the time when we inherit Ideas, we do so automatically, and that when we make selections, the reasons for those selections ultimately don't matter. But *because* we are human, I find it difficult to accept that *everything we do* is accidental and automatic, that *everything we do* ultimately doesn't matter. It makes me feel hollow inside. Surely there must be a point to human history.

There's something very familiar about the way I'm feeling. Plenty of people felt exactly this way in 1859, when On the Origin of Species was published. The cause of their unease was the book's most shocking proposal: that Life was on "autopilot." With Darwin's natural selection at the helm, no one needed to consider what Life did next, or what it had done, or even what it was doing. Consideration was not a consideration. The mindless, ceaseless engine of natural selection would take care of the flight plan—such as it was, because, according to Darwin, it was making it all up as it went along, and there was no real purpose to the flight anyway, no confirmed destination. Life was flying just because it could fly. And God, the universe's former pilot, was out of a job.

If hats and barns and every other artifact of humanity can self-design, if everything we hold dear can come about mindlessly, just as species do in nature, then *we* could be out of a job: cultural Life could evolve quite happily without any conscious participation on our part. As long as we host the Ideas, and compliantly pass them on to others, and select some Ideas in favor of others for whatever reason—in the grand scheme of things, the reasons themselves *don't matter*—the Force will always be strong in the noosphere, and human culture will come to possess all the characteristics of a capitalized Life. And if so, then this is who we are: smart but unquestioning Idea machines hosting an auto-evolving culture as it continues its pointless, almost imperceptible shuffle to nowhere.

How do you like that Idea? Not much? Neither do I. As a human being I'm affronted by the suggestion. Laying off God is one thing, but taking our species out of the picture? Well, that's something else. It's the ultimate in nihilism. It takes the value and purpose out of everything we do. Is there no space in this new world view for human genius, for "great leaps forward," for mad inventors and inspired thinkers? Can't we be consciously, actively, creatively employed somehow, driving cultural evolution, steering humanity?

Ads hums to himself and fiddles with the wheel for something to do. The Chrysler's in cruise again, so on a road as featureless as this, the car kinda drives itself. Tilted toward us, presented to us, are the lucky farms that sit on the mountain front. All of them have gambrel-roof barns. Even this far west, after all these miles, the barn raisers were still knocking out the same barn. Were they doing so because it was still, even here, the best barn for the job? Did they consider all the options? The farms here are small cattle farms, very unlike the huge cereal estates that bear these barns back east. Did the Bighorn barn raisers raise these barns consciously (as we'd like to believe), or did they do it accidentally/automatically (as we wouldn't), unthinkingly adopting their forefathers' barn designs in the same way that they unthinkingly adopted their forefathers' accents?

How can I tell? How can I hope to judge the extent to which the evolution of any Idea has veered to the left or right on the mindlessness-to-mindfulness spectrum over time?

Watchmaking

This situation has a precedent. In 1802 the British theologian William Paley put forward an argument in his book *Natural Theology* that he believed would finally put to bed the debate over the existence or nonexistence of God. It wasn't his Idea—it was a classic in teleological reasoning—but he brought it up to date by framing it as the "watchmaker analogy." It went like this: just as the complexity and marked design of a watch is evidence of the existence of a mindful watchmaker, so the complexity and marked design of nature is evidence of the existence of a mindful "nature maker": God.

On the face of it, it's a cracker, and it made a distinct impression on the mind of a young Christian theology student at Cambridge University named Charles Darwin. Darwin had always been a keen natural historian, and he interpreted Paley's book as a call to action. He decided to dedicate his time to documenting the detail of nature's complexity and design in order to qualify just how intricate the workings of God's watch were. He busied himself collecting varieties of organisms and rocks on Paley's behalf. He paid particular interest to the variation of individuals within a species: the tiny differences in the shells of barnacles, the gradations of leg length in stag beetles. The natural world was a watch of infinite involvedness, it seemed to the young Darwin, and he even accepted an offer to travel around the world on the HMS *Beagle* in order to sample some more of it.

But twenty-eight years later, after seeing the world, and to his great surprise, Darwin's conclusion was that Paley and Paley's watchmaker analogy were fundamentally wrong. In straining his eyes to look closer than anyone had before at the cogs and springs of the watch we call nature, Darwin had discovered how design might arise without a mindful maker, how an energetic but utterly mindless Life could indeed make the most beautiful and intricate watch imaginable. It all depended upon a witless repeating mechanism, natural selection. As Darwin wrote later in his autobiography:

> The old argument of design in nature, as given by Paley, which formerly seemed to me so conclusive, fails, now that the law of natural selection has been discovered. We can no longer argue that, for instance, the beautiful hinge of a bivalve shell must have been made by an intelligent being, like the hinge of a door by man. There seems to be no more design in the variability of organic beings and in the action of natural selection, than in the course which the wind blows.

Of course, he was right; but here I am, 150 years later, viewing the design of the hinge not only of a bivalve shell but also of a door as examples of natural selection's mindless works. Where Darwin's goggles suggested there is no nature maker, these goggles suggest there is no *humanity* maker. Where Darwin's said the biosphere was made accidentally, mine say the *noosphere* was made accidentally. Where Darwin said no mind was involved at all, what am I saying—that no mind was involved at all? How can I discover if these goggles are right?

How did Darwin discover that he was right? Answer: he set himself (and his goggles) a challenge. He reasoned that he could justifiably attack the rhetoric of William Paley's watchmaker analogy only if he could trace the gradual design through time of what he considered to be nature's watch: the eye. Like a watch, the eye is, on the face of it, impossible to make by accident. It's a device not only of remarkable utility and exquisite design, but it is also made up of many parts, none of which can work alone with any effect. How could the various components of the eye come together *without* a mindful intervention?

> To suppose that the eye . . . could have been formed by natural selection, seems, I freely confess, absurd in the highest possible degree. Yet reason tells me, that if numerous gradations from a perfect and complex eye to one very imperfect and simple, each grade being useful to its possessor, can be shown to exist . . . then the difficulty of believing that a perfect and complex eye could be formed by natural selection, though insuperable by our imagination, can hardly be considered real.

Numerous gradations, each grade being useful to its possessor—that's what Darwin needed to find. So he searched for them. He searched through the drawers and cabinets of Britain's natural history museums— he searched *through history*—and after some time, he succeeded in finding his gradually evolving eye stuck in the faces of organisms from all over the living world. From organs that consisted of nought but pits of light-sensitive cells to those even more complex than our own, and every tiny step in between. Darwin discovered his demonstration of a watch built by gradual, mindless modification. Where's mine?

What about the watch itself? That's an impossible *cultural* artifact. Can I find numerous gradations in *its* natural history? Well, yes. The starting point is to recognize that while Paley was right in assuming that a watch *is* evidence of the existence of a watchmaker, he was very wrong in assuming that a watch*maker* was the same thing as a watch *designer*. When a watchmaker sits down to put together a watch, she's not inventing the device from scratch; she's feeding off a long line of design work, constructing an example of a device that was gradually invented over thousands of years by thousands of different minds, tweak by tweak. Paley's comparison between a watchmaker and God was completely misleading. A watchmaker is not a mindful creator. A watchmaker is someone who arrives after a significant design journey has been completed and puts the evolved, inherited, accrued Ideas to good, practical use. Somewhere in Asia, at this very moment, an assembly line of robots is making watches. Each of them is performing a set task, building the watches bit by bit, just as genes construct eyes. "Making" is about performing copied routines, and copying is not a creative act. So the existence of a watch, while evidence of the existence of a watchmaker (or a line of robots), is not evidence of mindfulness per se.

Paley's blind assumption, one that everyone in his day shared, was that design had to be a mindful process. Darwin's great coup was to examine this assumption under a brighter light. *With enough time*, he revealed, the witless, repeating mechanism of natural selection can turn even design into a mindless act. With enough time (in his case, hundreds of millions of years), it can mindlessly invent something as complex as an eye. But has there been enough time for it to mindlessly invent a watch?

The history of design work in timekeeping is certainly long. Humans have been measuring the passing of time for at least four thousand years, and the evidence of numerous gradations in the design of these timekeeping devices, each grade being useful to its possessor, is also to be found in the drawers and cabinets of museums, albeit different museums. History is a tale of evolution, of gradations, of change through time. Open enough drawers and cabinets and the history of the timekeeping device will be laid out before you, from the first sundial through various water clocks, the invention of the ticking "escapement" mechanism, of spring-driven technology and the pendulum, to the use of electronics and quartz, to watches that Paley himself wouldn't have understood, watches that contain no clockwork parts at all! Every little shuffle step of this journey *could* have happened accidentally / automatically, mindlessly. But how do I know for sure?

There's a big difference between what Darwin was trying to do and what I am now faced with. Darwin was charged with discounting the involvement of one big mind, God's. Since the prevailing view in Britain at the time was that God had brought about the world in an act of Creation, all Darwin had to do to discount this (all he had to do!) was discount Creation. He did this by holding up the discovery of time—the ancient rocks full of fossils, the ancient designs in the jars of formaldehyde—to demonstrate that the world had not suddenly come into being, that it had enjoyed a substantial *natural* history.

My situation is different. We all accept that there has been a substantial human history. We also all accept that there has been an evolution of culture over time. My charge is not to discount the right-hand, mindful end of the spectrum, and hence conclude "it must have been the left-hand end, then." My charge is to work out exactly how much human history has wobbled about in the middle of the mindfulness-to-mindlessness spectrum. My charge is to discount, or otherwise, the varying involvement of not one big mind but billions of little minds.

Differently Dull Flip-books

You know those flip-books that you can get for children? They have a drawing toward the corner of each page, and by carefully bending the book back and securing it with your thumb, you can flip through the pages, and those drawings, revealing a little low-tech animation, aimed to amuse. Well, imagine a flip-book that contains a portrait of you at twenty years of age on the first page, a portrait of your mum (let's go with the maternal side) at twenty on the second page, your gran on the third, your great-gran on the fourth, and so on. In other words, imagine a flip-book with a portrait of each generation of your direct ancestry as a young adult on each successive page. How big is this flip-book? At a guess: two hundred billion pages. That's probably how many direct ancestors you have, two hundred billion, give or take a few billion, right back to the original living thing on Earth.

Can you imagine this flip-book? Good, let's flip!

Actually, we'll start by "jumping" through the book rather than flipping. Your ancestry flip-book is a hefty tome, and I doubt your thumb could take it from a cold start. It needs some exercise. So, open the book about a thousand pages in. You'll see a cavewoman. She looks almost the same as you (if she started looking through the jeans in The Gap next to you, you wouldn't bat an eyelid). Remember, for about two hundred thousand years, or ten thousand pages of the flip-book, our species hardly changes at all. So, let's jump back another nine thousand pages. Somewhere around that part of the book, there'll be a drawing of a special cavewoman. She's special because, if you managed to get hold of everybody's ancestry flip-book and turned to this page, or thereabouts, there she'd be, the same woman in every one of them. Her name is Mitochondrial Eve, and she is everyone's mother, the ultimate mother of all living humans.

From that page on, all our flip-books are identical (so you might as well give everybody his book back). As far as our species is concerned, after the ten thousandth page or thereabouts, you're flipping through the common ancestry flip-book of all *Homo sapiens*. So what lies beyond, in the other 199,999,990,000 pages? It's time to take big jumps. Another 20,000 pages, you'll see a woman with a big, ugly nose and a sloping forehead; you'd guess that she wouldn't have been a great conversationalist. Another 100,000 pages, you'll be able to say, "My, grandma, what a big long, protruding jaw you have!" Double your current progress into the flip-book and you'll be staring at a slender ape with nice eyes that sparkle. Double it again and you'll now be looking at something that resembles a chimpanzee. It's not a chimpanzee (because chimpanzees are modern creatures with their own 200-billion-page flip-books), but it is an ancestor we share with chimpanzees.

Leaping on . . . to page 2.5 million: Grandma is a monkey without a tail; on page 4 million, she's a strange squirrel-thing with a pointy nose; on page 40 million, a chubby rat; page 55 million, a stub-nosed monitor lizard!; page 70 million, a lungfish . . . and on and on, back through various incarnations of all our ancestors: fish-things and worm-things and sponges. And each time you triumphantly slap down the next 100 million pages and take a look at grandma, the animal looking up at you is hugely different and even weirder than the one before. And smaller. Grandma's shrinking! Pretty soon, after the worms, the drawings start to get so small you can hardly see them. You scramble around to find the plastic magnifying glass that you got last Christmas, but even then, the creatures fade to a dot. At about page 16 billion, squint as you will, all our grandmas have disappeared. With barely 8 percent of your flip-book browsed, and 184 billion pages still to go, you've lost sight of your ancestors; their portraits are just far too small to see.

Which is a shame, really, because if you're like me, you'd love to skip to the last page to find out how it ends . . . or begins; but you can't.

However, that's not the point. Time to stop jumping and start flipping. It doesn't matter where you choose to do it—at the ape end, the lizard bit, or in the middle of the fish—grab a couple of hundred pages, brace your thumb, and start to flip. What do you see? Virtually nothing of interest. The dullest flip-book ever made.

This is how evolution works at the left-hand end of the mindlessness-to-mindfulness spectrum. It's an almost imperceptible shuffle. Natural selection may have granted each of our ancestors the permission to work on their "self-design," but it did so while imposing the most torturous work restrictions imaginable. Each and every one of our ancestors was permitted to work only on the design of *their* immediate ancestor. They couldn't steal traits from others in their generation. There was no "blue sky thinking." They couldn't "start from scratch." Their only option was to tweak the characteristics of the individual who came before them: a bit here, a bit there. If they inherited design faults or redundant features designed for a very different past environment—tough! If they lacked the materials to achieve the ideal design—bad luck! The whole thing was a botch, a make-do-and-mend. It's a stupid way to approach a design project. Mindless, in fact.

If cultural evolution was a 100 percent mindless Darwinian process, confined to the extreme left-hand end of the mindlessness-to-mindfulness spectrum, that's what you'd expect to see throughout all cultural history: a botchy, gradual "descent by modification." Every Idea would be confined to an almost imperceptible shuffle, prototype after prototype, each dealing with the immediate concerns of the time, impeded in its options by its own history of design. If cultural evolution were in fact fully mindless, then there could be no original Ideas; each "new" Idea could only ever consist of its immediate ancestor *plus a few tweaks*. Admittedly, some of those tweaks could be revolutionary—the same thing happens in nature—but for the vast majority of the time, progress would be achingly slow. Just like me and you, every Idea would have a very dull ancestry flip-book.

However, if our minds *were* involved intentionally/consciously in the evolution of Ideas, the flip-books would be more exciting. Indeed, it's a proportional relationship: the more mindful an Idea's evolution, the more exciting its flip-book would be. You'd see great leaps between pages—wildly different portraits resulting from acts of "creativity," right-hand-end action—the sort of thing I've called artificial creation in the Life of the biosphere. The evolution of an Idea would not need to be blind to the future; it could evolve traits that planned ahead rather than simply reacted to the present. Ultimately, there could be acts of genuine "natural" creation, brand-new Ideas without ancestries, Ideas that arrived "out of nowhere" ready-made to fit a vacant niche in the noosphere. If human minds were routinely, mindfully involved in cultural evolution, the evolution of Ideas would be extremely exciting.

So which is it? The barns that fly past as we continue on the freeway toward Montana must have their own ancestry flip-books. Are they really as dull as ours? Are there no great leaps of creativity, no bounds of human ingenuity to scupper the claims of the cultural Darwinists? Is there nothing in the history of barns that would smash these goggles once and for all? What about the sudden appearance of gambrel roofs on barns across the States in the latter half of the nineteenth century—a new barn "out of nowhere"? Wasn't that evidence of human genius, the right-hand end of the spectrum, the mindfulness of a cultural creation?

Auto-barn

At our coffee stop in Sheridan, Wyoming, I went online and discovered a peculiar fact: long before a gambrel roof topped a barn, it topped a house. The oldest existing timber frame building in the United States, the Fairbanks House in Dedham, Massachusetts, has two gambrel-roof extensions dating from the 1660s. This means that rural Americans were putting this ingenious roof design on their houses for 150 years before they got the Idea to incorporate it in their barns. Until that point, they must have (automatically) built English, Dutch, or German barns as their forefathers had always (automatically) done.

And there's more. According to one theory, the gambrel roof wasn't even an American invention; it was brought over by English and Dutch minds in the mid-seventeenth century, and they, in turn, got the Idea from the timber houses they witnessed upon first setting foot in Southeast Asia the century before that. In that part of the world, gambrel roofs had been placed (automatically) on houses for unknown generations. So the gambrel-roof Idea didn't come from "out of nowhere"; it came from "out of somewhere else." It was invented way back in the Stone Age—another age—plenty of time for the Idea to have evolved at a shuffle; there are no leaps and bounds in evidence here.

So it's likely that the gambrel-roof barn has just as dull an ancestry flip-book as we do, that it evolved as mindlessly, as accidentally, as automatically as an accent, or a cowboy hat, or a watch.

But sitting here in the Chrysler, in cruise, I can't resist the conviction that not every Idea in the noosphere shuffles along in cruise, as we do in our Chrysler. Perhaps it's because I've taken my new-world-view goggles off and have thus returned to my former assumption that we humans *are* at the steering wheel, or perhaps it's because of a memory that I have from when I was younger, from my first trip to America. I'm ten years of age and I have my nose pressed up against the big windows of a terminal in JFK airport. I'm gazing at the distant bumpy line of Manhattan. Since I'm from southwest England, I've never seen a jungle of skyscrapers before. My dad sits down on the carpet next to me. "That's America," he says, "a whole continent of cities, roads, and farms—all of it built in less than five hundred years. Amazing, isn't it? If you started from scratch now, with all the planners, architects, builders, machinery, and money the world has to offer, you couldn't imagine being able to build America from scratch in less than five hundred years. You just couldn't get it done."

He was right! If humanity is on autopilot, then how does it manage to perform such dazzling acrobatic displays? The human world doesn't shuffle along; it travels at sickening speeds; its movement is eminently perceptible. Surely the pace of change on this continent cannot be put down solely to natural selection; there must be *minds at work*. There just isn't enough history for 100 percent mindlessness—certainly not in America. There has to be a space in this new world view for our human genius.

Before I arrive at the "Tepee Capital of the World," I'm going to set myself (and these goggles) a challenge: Can I reduce America's history to a mindless brawl between shuffling, gradually evolving Ideas? Can I gaze again upon the complexity and marked design of America and have the gall to conclude that no one made it, that even though it rose from dust in less than five hundred years, there were no, and are no, "America makers"?

Goggles back on.

8

How the West Was Won I: Finding the Edges

Hear the Herd?

It's like I've gone deaf. We've gotten to southeastern Montana and I'm standing on a small hill. The landscape is enormous in all directions. It hangs out there just beyond reach, but I can't seem to hear it; there's only a breeze that whispers and flicks at my eyes. The midafternoon sun is hot on my cheeks and blinding; tears roll as I try to stare south. There's a strong scent of pollen, desiccated chlorophyll, and the heated silica edges of a billion billion blades of grass.

One hundred and fifty years ago this spot would not have been so silent. It would have been raucous with an animal that once commanded the Great Plains. The buffalo, or more properly the American bison

(*Bison bison**), is the largest land animal in the Americas. A bull buffalo can be six feet tall, ten feet long, and weigh more than twenty-five hundred pounds. They were the chief lawn mowers of the plains. Buffalo were so efficient at keeping all this grass cropped that early trappers soon learned that the chief barrier to their work on the plains was not the resistance of local Indians but the task of finding shoots long enough for their horses to eat.

For countless years, Native American tribes living on and around the edges of the plains sent bands of hunters into the basin now laid out before me. It was always one of the best places in North America to hunt buffalo. Any herd wanting to move south in the summer to pick up the green shoots that emerge after the seasonal fires, or north in the winter to find the spots where the Chinook wind strips the snow off the turf, would be channeled between the Bighorns and the Black Hills—over this exact spot.

The hunting parties would set up their nomadic camp of tepees in the river valley (below me), where freshwater was available and the scattered copses of deep green cottonwood trees afforded them a little shelter from the wind and the sun. Just like the grasses, and indeed the buffalo, we humans are engaged in a constant battle with these elements.

The hunters knew from experience when the buffalo would come through the area, but they would send scouts out to locate a herd's precise position in advance of the hunt. The scouts would get to high ground to spot the herd, but they would also listen for it on the wind: it clattered like a giant decrepit machine. In spring, boisterous buffalo cows bellowed out orders to one another. In summer, once the bulls had joined the herds, a bovine metronome of barks and bays coughed all day and night, splitting the mass up into guarded harems. If shouts weren't enough, the bulls would fight, cracking heads and sending a ripple of fearful bleats through the neighboring herd. In fall, the shaggy bulls left the herd to walk the plains in small groups, and the cows became more placid, but the air would be filled with the whining of weanlings, first-year calves, protesting the reluctance of their mothers to offer more milk. When winter threatened, the wolf packs were busy, and the herd

---

*I'm not going mad; that's the Latin name for the species *bison* in the genus *Bison*. The American plains buffalo is called *Bison bison bison*—the subspecies *bison* of the species *bison* in the genus *Bison*. And I'm worried about finding an original *Idea!*

would stampede, giving itself away with a storm of dust and heat in the frozen air and a thunder of hooves.

Buffalo are rambunctious beasts, but because of the scale of the plains, even a million noisy buffalo had to be scouted out. When the scouts had detected the herd, the village in the valley would spring into action. Each tribe was utterly dependent on buffalo meat to sustain it through the hard, hard middle-of-a-continent winter. A good hunt would mean that the tribe would feed well throughout the season. A failed hunt would mean starvation for some, and a net population loss in the tribe that could leave it unable to defend its hunting lands in the coming spring. Tribes could disappear altogether if their prowess on the plains was poor. Some members would starve, some would wander off to join relatives elsewhere, and some would succumb to the war parties of tribes that had fared better the previous year. It was a ruthless form of cultural natural selection, and one that was effective at honing one of the most iconic human ecotypes: the Plains Indian.

But every year in the spring, one tribe or other would set up a village of tepees in the valley below me, from some year unknown to 1876, when history dictated that the Plains Indian culture would draw its last breath, right here on this spot.

Trail and Error

The last breath of the Plains Indian culture had been an inevitability ever since the European culture took its first breath on American soil hundreds of years before. Once the Europeans had found America's edges, it was only a matter of time, everyone unfailingly agreed, before the new people of the New World filled in the middle and forced the native people to "melt and vanish before the advancing waves of Anglo-American power," as the great historian of the American West Francis Parkman put it. Since this outcome was inevitable, it must, the new Americans concluded, be destiny. And since it must be destiny, it was, they deduced, Divine Will.

In a December 1845 article designed to inspire Americans to grab yet more of the continent, crowd-pleasing journalist John O'Sullivan provided a catchphrase for this sentiment of divine will: "and that claim," he declared, "is by the right of our manifest destiny to overspread and to possess the whole of the continent which Providence has given us for the development of the great experiment of liberty and federated self-government entrusted to us." Americans were on a mission to cover the whole landmass, and they had God on their side.

At the time O'Sullivan was writing his article, the United States was only about two thirds its current size. It had stretched over the continent from the Atlantic to the Rockies, securing the former Republic of Texas as its twentieth state that year, but its western border didn't yet have a beach. Between it and the coast were the unexplored wildernesses of "Oregon Country," claimed by the British, and "Alta California," the Mexican province. Fulfilling the manifest destiny of the young nation could mean only one thing: a policy of mass illegal emigration into the unknown.

At that time, the only Americans to have ventured anywhere close to the unknown were fur trappers. Crisscrossing watersheds on the lookout for the valuable beaver pelt, approximately three thousand of these "mountain men" performed their random surveys of the continent for twenty years until, in the early 1840s, the bottom fell out of the beaver. The public will to migrate west into country of which the trappers had some knowledge couldn't have come at a better time for these businessmen. Turning their ponies around and giddyupping them west again, the trappers led wagon train after wagon train of hopeful migrants, their lives packed in canvas-wrapped Studebakers,* up onto the high plains. But they soon discovered that they had a problem: their pony trails, bobbing in and out of countless valleys, were useful for finding beaver, but of little use when fulfilling manifest destiny. Their need now was for a direct route to the Promised Land, with a passage wide and gentle enough for a wagon train to conquer. Their only choice was to take on, by trial and error, the labyrinthine terrain of the Rockies and the Great Basin beyond. There was no guessing where the best high passes or the most fruitful valley floors were hidden; the vanguard just took a bearing west and tried their luck.

Progress was tough. They found themselves trapped in box canyons, meeting escarpment walls, and tracking for miles along rivers with no safe crossing. They doubled back. They went around. They battled over the top. It was a shuffle. Each new position was only ever the previous position plus a tweak. They were impeded by their own history in the

---

*The Studebaker family made wagons before they made cars.

landscape: the choices they had already made or, rather, that had been made for them by the characteristics of the environment itself. But how else do you cross a landscape you don't yet know? Journeys in strange landscapes will always be endeavors driven more by luck than skill. Only once a destination is reached can you really know the way. So instead of finding a trail, the trail found them. It kinda designed itself.

The mindless ants that inhabit the plains and river valleys through which the wagon trains wobbled find the same thing. They blaze trails to the Promised Land in a very similar fashion. Ant colonies forage by sending scouts to span out at random over the neighboring landscape. Once a scout is lucky enough to find the "Promised Land"—in the form, perhaps, of an apple core cast out of a wagon—it will bite off as much as possible and then follow its own trail, no matter how indirect it is, back to the nest. This makes sense; after all, it's the only route home the ant knows. Well, it doesn't actually "know" the route; it doesn't remember it. It finds its way home by following a path of the scent markers, pheromones, it exuded on its outbound journey. Because the ant scout travels the same path twice, once outbound and once homebound, and because it never stops exuding pheromones, by the time it returns to the nest, the trail to the Promised Land is marked with two doses of pheromones, which is exactly double the dose of all the other, unfruitful trails.

If any other scout ant, on its random survey of the territory, happens upon this double-dosed trail, it will take it—guided by an irresistible urge always to follow the trail with the most pheromones. In time, other ants will run into the apple core trail at different points. They will likewise find the path to the Promised Land, grab a bite, and head back to the nest via the route with the most pheromones, laying down more and more pheromones as they go. However—and this is the clever bit—because there are lots of ants in an ant colony, and because randomly motivated ants will join the trail to the Promised Land at many different points, over time the trail will inevitably "self-optimize." By this I mean that ultimately the ant colony will discover, even though it has no mind, the shortest route to and from the apple core. This is due to one critical factor: pheromones evaporate. The longer the route to the core, the longer it takes an ant to travel it and the more time there is for the pheromones on the trail to disappear. So the trail with the most pheromones on it will always be the one that covers the distance in the shortest time. Simply by following a few lines of algorithmic programming in their heads, the mindless ant colony will discover the shortest route to the core. With trial and error, unconscious communication, and a lot of scouts, ants will inevitably blaze the ultimate trail to the Promised Land. The way they operate in the environment means that they are destined to do so.

If you look at a map in a U.S. history book, the trails to Oregon, the Great Basin, and California look like direct routes to a Promised Land, but those maps were drawn up only after the trails had been completed. At the time—close up, in the middle of the continent, in the middle of the century—the designing of those trails involved people getting lost in huge valleys, wagons washing away in rivers, Jim Bridger stumbling upon a new pass, and *American* illegal immigrants trying to find the best routes to avoid the *Mexican* border police. It involved lots of scouts venturing into unknown landscapes. And if these immigrants happened upon wagon tracks, they tended to take them: the deeper the ruts, the more likely it was the tracks led to the Promised Land. And since wagon tracks, like ant pheromones, will fade and disappear with time, perhaps it's not too fanciful to imagine that with enough trial and error and enough unconscious communication and enough scouts, the routes that we now see in the history books, the ultimate trails to the Promised Land, were destined to become manifest, just as O'Sullivan ordained.

The Southern Herd

In 1871, Colonel Richard Irving Dodge (after whom Dodge City was named) took a trip from Fort Dodge (named after a completely different Dodge) along the Arkansas River, Kansas, in a light wagon. A few miles from the fort, Dodge entered a herd of buffalo traveling across his path. Four hours later he came out the other end. Buffalo had filled his horizon for twenty-five miles of the thirty-four-mile trip. Since the buffalo were moving the entire time, it suggested* that the herd must have been at least twenty-five miles square, which, accounting for variances in buffalo density, would imply that four million animals moved south through Kansas that day.

*To one E. T. Seton in 1927.

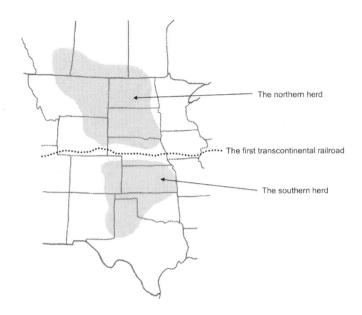

The splitting of the buffalo herd upon the completion of the first transcontinental railroad.

Colonel Dodge had happened upon the great "southern herd." For millennia, the buffalo had wandered liberally over the plains in herds that continually merged and split like slime molds. But the forging of the Oregon Trail in the 1940s sliced the native herd in two. With each passing wagon train, more and more buffalo would fall: the pioneers needed the meat; the trappers needed the buffalo pelt, or "robes." When gold was found in California in 1849, the rush turned the trail into a thoroughfare, and the buffalo learned to keep away. Because it was the ultimate route through the Rockies, the trail was also the natural choice for railroad engineers laying track through the Wild West in the 1860s. By the time a "golden" spike was driven in to join the Central Pacific and Union Pacific railroads and open the first continuous railroad between eastern and western United States in 1869, the American buffalo herd had become permanently fragmented into two populations: a northern herd on one side of the tracks, and a southern herd on the other. On that day in 1871 Colonel Dodge met the herd on the southern side. As it turned out, they were on the wrong side of the tracks.

Bursting forward from the new railhead towns and onto the southern plains almost as soon as those towns were up and running was a new brand of American: the commercial buffalo hunter. Even as the Indian tribes were being pushed back and onto reservations, these people busied themselves killing their prey in the tribes' wake. They would chase down small groups of buffalo on horseback with Colt revolvers. It was dangerous work. As the herd stampeded, the grassland would turn to dust, and horse and buffalo hooves would snare in prairie dog holes, catapulting riders into the throng. After the stampede, the hunters would strip the robes, butcher the carcasses, and drag their goods to the nearest rail station market. They'd get very little for the coats and even less for the meat, but these were men trying to live in a nation recovering from a civil war. Many were ex-Confederates looking for a corner of this new Union nation in which to start again, their minds bloodied with the horrors of the battlefield, their eyes skilled at picking targets. Killing buffalo in the Wild West for whiskey money was a lifestyle that suited them. And they had endless energy for it. Within a year they had cleared the buffalo from the country bordering the railroad. Each half of the divided herd responded by drifting north and south, away from the Union Pacific.

The buffalo could outrun one railroad, but not, it turned out, three. A year later, a second railroad, the Kansas Pacific, lit up a chain of towns to the south of the Union Pacific. The new railroad shot straight through the heart of the southern herd, an injury the buffalo would not recover from. Even as Colonel Dodge clippity-clopped through the head of this herd, the first of the Kansas buffalo hunters were making their way down from the new railroad towns to shoot at the herd's rear.

In the years 1871–1874, buffalo hunting became industrialized. Building on their previous season, and recruiting professional hunters from across the continent, the buffalo prospectors developed a new style of killing: still-hunting. Rather than risk your life in a stampede, you would locate a small band of buffalo, say, one hidden in a gully away from the main herd, and creep up to them on a rise. Lying low, you would pick them off sniper-like, but in a certain order. Buffalo herds are led by old cows, and they are the brains of the herd. They decide when the herd will move and where it will move to. If you shoot a herd in the brain, the herd will stand there, dead. So you watch the group for a while and work out which buffalo is the brain. Then you shoot it. You don't want it to drop suddenly and spook the rest, so you shoot it just back from the forelimb and about a foot above the brisket, in the lungs. The noise will startle the herd, and they will spy the puff of smoke on the hill, but with their brain standing still, they will stand, too. Shortly after the shot, blood will trickle from the old cow's nostrils. A few minutes later, she will shuffle on her feet, try to keep upright, then she'll sway, lose her balance, and fall. The others will gather around her, wide-eyed in bewilderment, awaiting instruction. After a while another old cow will take the reins and begin to move. But as soon as you spot which cow has become the new brain and has started to lead the herd away—bang!—you stop her in her tracks, and the herd numbs again.

This method of hunting is possible only with a long-range, breech-loading rifle. Unfortunately for the buffalo, you need the same sort of gun to kill your fellow countrymen from a distance. So, like the men who did the shooting, and the destitution that had led them to shoot, the armaments of this war against the buffalo were born of the recent Civil War. Losing sales after the war, Sharps, the rifle company, was quick to adapt its bestseller for the new killing grounds by launching large-bore versions that could puncture the hide of a bull buffalo from 250 yards.

Once the slow business of finishing off all the old cows is over, the remainder of the group will more than likely just stand there and wait to be shot. The limiting factor on the time it takes to finish them off, then, is not the accuracy of the hunter—any hit will do—but the overheating of his rifle. The black powder Sharps got so hot that you had to rest them between shots. But in the winter, when the buffalo robes were at their thickest and most desirable, you could rely on the snow on the ground to cool down the barrel and hurry up the chore. A good still-hunter could take one hundred buffalo in an hour.

As buffalo products started flowing east on the trains, the appetite of the urban population of the United States would wax and wane for different buffalo parts. At points, the fickle fancies of city consumers, and a lag in the hunters' response to them, would turn parts of the buffalo into gold. One month, the horns were the things to collect; the next, the tongues were worth more. Because the general view was that the buffalo herd was inexhaustible, no one set up plants for the efficient processing of the dead. The lone still-hunters would simply rip off the parts they needed and leave the rest to the coyotes.

There were not enough coyotes, however. In the fall of 1873, William Blackmore traveled along the same bank of the Arkansas River that Colonel Dodge had journeyed along two years before, but instead of wading through a living herd, he found a ghost herd: "there was a continuous line of putrescent carcasses, so that the air was rendered pestilential and offensive to the last degree. The hunters had formed a line of camps along the banks of the river, and had shot down the buffalo, night and morning, as they came to drink."

The Nature of Panic

At the same time, the ruin of a different kind of animal, the investment banker, was occurring in New York, also as a result of the railways. Between 1861 and 1865, America's manifest destiny was put on hold, but as soon as the Civil War was done, the universal mood across the States was to get on and realize that destiny. Railroad construction right through the wilds seemed the ultimate manifestation of this, and postwar investment came easily. Thirty-five thousand miles of track was laid toward the Pacific between the war and 1873, involving millions of dollars of financing, before someone asked when a return on their investment was destined to become manifest. The answer can't have been comforting, because it caused a financial panic. There was a mad scramble. Investments collapsed across the board. The New York Stock Exchange closed for ten days in an effort to control the madness, but it was too late. One quarter of all the railroad operators went bankrupt. One of the biggest banks at that time, Jay Cooke and Co., buckled in the aftermath, along with the immediate hopes for a continuation of the Northern Pacific Railway (the last of the three buffalo killers), which then stopped in its tracks.

Perhaps the southern herd could have been saved if this hadn't happened. Had the Northern Pacific reached Montana the following year as promised, the hunt for robes may have shifted to the still-huge northern herd in 1874, taking the pressure off the few buffalo that were left in the South. But the panic did happen, and as a result, the tragedy of what ecologists call "scramble competition" fell upon the southern buffalo herd. Scramble competition happens often in nature, wherever a growing number of consumers have equal access to a finite resource. It always ends in a pile of dead bodies lying on an exhausted stock—a mindless tragedy because nature is mindless. And so, it appears, are we. We have gold rushes and runs on banks, we strip rain forests and overfish the seas, and we hunted the southern buffalo herd until it disappeared. Civilized humans, with all our intelligence, are as subordinate to the rules of nature as a buffalo herd without a brain.

Even late in the winter of 1873 the hunters thought that there would be no end to the buffalo in the south. Then, suddenly, the desperation of the situation—the exhaustion of the stock—fell into view. But the rarity of buffalo didn't stop the hunters. The rarity of the buffalo meant that each new robe or tongue was worth more than the last. The failure of the Northern Pacific to seek out the northern herd meant that the price of buffalo rose as the herd fractured.

Natural selection finally whirred, and the southern buffalo stock was reduced to only the wariest animals. When the old cows fell, these buffalo retained their uneasy feeling of danger, even after a new cow took charge. They became their own brains, and dropped into first a trot and then a gallop as more shots rang out. Of this subset, there was further selection against animals that did not instinctively head for trees or mountains. The still-hunters dropped their rifles, mounted their horses, and headed off across the grass after them. They followed the buffalo tracks onto the high plains of Texas. They hung around watering holes and salt licks. They pursued every buffalo that was seen by drifters until the only buffalo left in the south were the very few that remained unseen.

Then, in 1880, with the financial panic over, the Northern Pacific Railway garnered enough cash to drive into Montana, and the hunters scrambled for the northern herd.

The Northern Herd

At this point in history the northern buffalo herd was vaster and more dispersed than the southern had ever been. It also had winter on its side. The northern winters were so tough that they drove off the hunters and dramatically slowed down the killing for a few months each year. But these subtle gains were neutralized by the fact that the buffalo hunt was now a fully industrialized operation. The hunters now knew exactly how to kill buffalo best; they'd learned. Also, this time they could recruit the assistance of the northern Plains Indians, who, destined to live on their fruitless reservations and aware that the loss of their traditional prey was also destined, soon decided to make hay while the sun shone and join the whites on the last great buffalo hunt. Together they hunted the northern herd for three years, and that was all it took to extinguish the rest of the American plains buffalo.

In 1886, William Hornaday of the National Museum in Washington was charged with discovering what was left of the species. He ventured out onto the plains of Montana and asked trappers, pioneers, and Indians if they'd seen any standing buffalo. Even all those years later, the bones of buffalo littered the grasslands, and Hornaday witnessed the pathetic toil of ex-hunters, with nothing to hunt, trying to make ends meet by collecting the bones of their victims. Apart from the clatter of bone carts, the plains were silent all around him.

After months of being told that there were no buffalo left in Montana, Hornaday located a group in the badlands where the Yellowstone and Missouri rivers meet. Almost all of them had bullets lodged in their hides. He estimated that there were about forty. He shot twenty-five of them. He needed them for the museum—now that they were so rare. This is how he described the encounter with the largest bull:

> When he saw me coming he got upon his feet and ran a short distance, but was easily overtaken. He then stood at bay, and halting within 30 yards of him I enjoyed the rare opportunity of studying a live bull buffalo of the largest size on foot on his native heath. I even made an outline sketch of him in my note-book. Having studied his form and outlines as much as was really necessary, I gave him a final shot through the lungs, which soon ended his career.

Hornaday's work brought the definitive audit of the species after the great scramble. The southern herd amounted to 25 in Texas, 20 in the mountains of Colorado, and 26 in southern Wyoming. The northern herd was 10 in Montana (5 more of Hornaday's group were killed when cowboys learned they were there) and 4 in Dakota. In northern Alberta, he heard of the existence of 550, but in fact these were all that remained of the wood buffalo, a larger subspecies that had received the same treatment at the hand of civilized humankind as its cousin on the plains. With 256 in captivity and 200 protected in Yellowstone National Park, Hornaday estimated that at the close of the nineteenth century there were just over 1,000 American plains buffalo left.

The parable of the buffalo has this to say: humans can be mindless, too. We're smart; we can (accidentally) come up with clever ways of hunting, efficient rifles, and trains to take the trophies away, but our intelligence is just a front. Behind those sparkling eyes we can be utterly mindless. Hornaday summed up the episode as follows: "Probably never before in the history of the world, until civilized man came in contact with the buffalo, did whole armies of men march out in true military style, with officers, flags, chaplains, and rules of war, and make war on wild animals. No wonder the buffalo has been exterminated."

9

How the West Was Won II:
June 25, 1876

Culture's Last Stand

In June 1876, between the two wars against the buffalo, an army of men marched out, again in true military style, with officers, flags, chaplains, rules of war, and Lt. Col. George Armstrong Custer, to make war on another living thing: the Plains Indian culture. The way of life of the Plains Indian had already been decimated. The demise of the buffalo was of course significant, but even if the buffalo herds had survived, the Plains Indian culture would not have. The tribes had been forbidden from continuing their nomadic lifestyle and placed on reservations. They were given food and clothes rations and canvas to make tepees, since buffalo hides would soon be unavailable, and told to cease the wild hunt and start tilling the grasslands. Their culture was collapsing just like the markets in New York.

While many Plains Indians reluctantly complied with the rulings from Washington, some did not, and for decades up and down the Great Plains, tribes had been clashing with the U.S. Cavalry. The retribution of white America upon this affront to their manifest destiny was brutal, and by 1876 only one renegade band remained wild on the plains, a band led by the Lakota Sioux holy man Sitting Bull. From the creation of the Great Sioux Reservation in 1868, Sitting Bull had refused to follow his kin and settle. He and his companions continued to hunt the northern herd and made their struggle political by attacking any wagon trains heading for the West, sacking forts and sending the surveying missions from the Northern Pacific Railway packing. Sitting Bull claimed the country's destiny for *his* people. His resistance didn't become a serious problem until 1876.

After the Panic of 1873, the United States fell into a depression, and unemployment went through the roof. Then George Armstrong Custer found gold on his expedition to the Black Hills, deep within the Great Sioux Reservation. Before he'd even returned to Fort Abraham Lincoln, the newspapers of the East were already declaring that Custer had saved the country from bankruptcy. Indifferent to the protests of the Sioux, thousands of white prospectors started making their way illegally onto the reservation and into the sacred Black Hills. President Grant sent an army to stop the trespassing, but in the fall of 1875, gold was again discovered, this time in Deadwood Gulch. Everyone knew that by spring the Black Hills would be swarming with prospectors, all scrambling to get at the gold.

Attempts were made to buy the Black Hills from the Sioux, but the Sioux chiefs refused. Desperate to curb the economic depression, Grant agreed to withdraw the army from the reservation, and the gold rush got the green light. At the same time, to keep some form of order, he commanded that all Sioux return to their settlements within the reservation—away from the Black Hills. In February 1876 he certified that any Sioux not residing on reservations could be considered "hostile." This certification effectively gave the military a legal right to hunt down Sitting Bull.

That spring, Sitting Bull's village of tepees sat on the Little Bighorn River in southeastern Montana, slap bang in the middle of the buffalo runway (just below where I now stand, experiencing the sensation of deafness). Women and girls made tepees and clothes from buffalo hide, boiled buffalo bones for glue, and made pemmican. The boys practiced shooting arrows and guarded the spare horses. The men rode out onto the plains and chased the buffalo herd. It was to be the last time the culture of the Plains Indian would live in its natural environment.

As the days went on, the village gained tepees. Sitting Bull's loyal band of eight hundred Lakota Sioux were joined by others intent on defying Grant's orders: hundreds of Sioux from across the Dakotas, and members from the Cheyenne and Arapaho tribes. In the past, the Sioux had clashed with both these neighbors, but Sitting Bull knew that they were all now friends, united as they were against a greater foe: the new world they were being forced to enter, a world in which their holy land was dug up for metal; a world in which their men had to stop hunting and join the women tilling the soil; a world in which their buffalo-hide tepees, their homes since the beginning of time, were about to rot and disappear forever. In the spring of 1876 these people joined Sitting Bull to hunt buffalo for their summer hides as they had always done. They would rather die in this old world than live in the new. By June 24, 1876, the village comprised a thousand tepees.

At sunrise the next day, Custer squinted from a vantage point in the Wolf Mountains (the line of low speckled hills to my east). His Crow scouts were claiming that they had found the village of hostiles, and that if Custer looked carefully he could see it in the valley, fifteen miles away. The scouts said that it was the largest tepee village they had ever seen. Custer peered into the valley on the plains, but his eyes weren't good enough. The Crow told him to look for maggots wriggling on the hills above the cottonwood trees—these were horses, in their thousands. Custer couldn't see them. He had been told that there were only eight hundred Indians with Sitting Bull. The Seventh Cavalry under his command numbered more than six hundred, and he considered his own presence worth a few more men. He decided to attack the following morning. Had he gone through with this plan, he would have been fighting alongside the troops of General Terry and Colonel Gibbon, who arrived in the valley on the twenty-sixth, and the outcome might have been very different. But an accident changed his course. Later that morning, a group of Sioux (who were actually making their way back to their reservation loaded with fresh buffalo hides) spied Custer's forces, and Custer's forces spied the group of Sioux. Custer reasoned that, with his position revealed, Sitting Bull would flee the Little Bighorn valley, that the Indians would scatter across the plains and into the Bighorn Mountains, and that the army would spend the whole summer hunting them down. Custer knew how quickly a tepee village could be collapsed and moved, so he changed his plans. He decided to attack Sitting Bull that afternoon.

Getting to the Phone

At exactly that moment, oblivious to the national disaster unfolding seventeen hundred miles to the west, thirty-five thousand Americans were in a park in Philadelphia wandering among the exhibits of the World's Fair. The World's Fair had become a fixture on the European calendar, a grand exposition—we call it Expo today—designed to show off the greatest achievements of the sciences and arts from around the world. This year, 1876, was the first year in which the fair was held outside Europe, and the United States relished the opportunity to show the Europeans just how far they had come in the one hundred years since the signing of the Declaration of Independence. They called it the Centennial Exposition, and made certain that it was a success. The total site encompassed 300 acres. The Main Exhibition Hall was then the largest building on Earth: over 21 acres housing 13,700 different designs of furniture, clothing, and household objects. The second largest building on Earth, Machinery Hall, stood next door, rammed full with pistons, turbines, and engines. Horticultural Hall was a soaring palace of glass, home to an indoor jungle lit at night by chandeliers. There were two observation towers, one of which had a steam-powered elevator; there was a coal-powered monorail that bridged a ravine within the park to transport fairgoers to a classy restaurant; and the right arm and torch of the Statue of Liberty were on display in an effort to rally funds to complete the rest of the famous gift from France. In all, there were two hundred buildings laid out in Fairmount Park. More than ten million people came through the gates in the six months the fair was open (wearing top hats, derbies, and bonnets rather than cowboy hats), one fifth of all the Americans on Earth.

Standing at one end of Machinery Hall that morning, hatless but encumbered by his stuffy, formal suit, a twenty-nine-year-old Scot was hurriedly rigging up his exhibit in front of a growing and growingly impatient crowd. He'd intended it to be in place when the fair opened in May, but his invention had not been ready for public demonstration at that time. Today, he was hoping, it was.

Alexander Graham Bell had had a busy twelve months. The previous June, his attempts to create a harmonic telegraph, a telegraph machine that could transmit multiple messages along a single wire, had taken an unexpected turn when his co-experimenter, Thomas Watson, had (accidentally) twanged one of the reeds at the transmitter end of the prototype. Standing next to the receiver, Bell had heard the various tones of the twang. What he heard immediately suggested to him that a single reed could be used to transmit not just a Morse code telegraph but all the complexities of human speech.

This may sound like a "great leap," a prime example of human ingenuity, but Bell's life had in fact been one long shuffle toward that exact revelation; he was uniquely attuned to host this "new" Idea. When he was twelve, growing up in Scotland, Bell's mother started to go deaf. Her gradual descent into silence forced the young Bell gradually to adjust his methods of communication. First he learned how to speak in clear, modulated tones so that she had a better chance of understanding him. Next he would talk directly onto her forehead in order that she might pick up his vibrations. Once she was entirely deaf, they both learned a finger language so that he could sit by her side and tap out the conversations of the people around them.

But this was just the start of his interest in the physics of communication. Bell's father, two uncles, and grandfather were all experts in the field of elocution and speech therapy. His father's particular expertise was teaching the deaf how to speak. This entailed securing a detailed knowledge of the relative positions of the larynx, throat, tongue, and lips during the formation of vowel and consonant sounds. Bell's father taught him all these details during Alexander's teenage years, so that by the time he was eighteen, Bell was qualified to teach the deaf himself.

Bell's first foray into experimentation on acoustics occurred when, as a teenager, he teamed up with his brother to build a model head with movable lips that could be posed and made to pronounce syllables when bellows were operated at the back of the throat. Understanding that the sound of all speech stems from the vibration of cords at the base of the throat, he moved on to experimenting with tuning forks, finding that if they were supplied with electricity, he could control their resonance and make them emit vowel sounds. He reasoned that with enough knowledge of electricity, he could manufacture any imaginable sound.

In 1870 his family moved to North America, and Bell became involved in teaching his father's techniques for speech recovery to deaf students in Boston. Every night after work and into the early hours, Bell continued his electro-acoustic experiments. He built all sorts of gadgets and gizmos to transmit sounds. He patched electromagnets onto old Morse telegraphs. He built a piano that could, by means of electricity, transmit its music to a distant point. He constructed an apparatus with a drum made from cow intestine that would vibrate when an intermittent electrical charge arrived. So when Bell heard the overtones of that single reed twanged accidentally by Watson in 1875, what was a giant leap for humankind was merely a small step for Bell.

At the time he took that step, Bell was working on the harmonic telegraph for two wealthy patrons, Gardiner Hubbard and Thomas Sanders. When he told them of his new experiments to transmit *voices* rather than Morse code over the telegraph, they were initially not so keen on the distraction. The harmonic telegraph was the Holy Grail of 1875. The telegraph network was regarded as the "nation's nervous system" at that time, and the financial rewards of sending many telegraphs down one wire, were potentially astronomical. Several key inventors, including Thomas Edison and Elisha Gray, were busy working on one; there was a race on. So Bell had to hurry to invent a harmonic telegraph and submit a patent before he could go on to conquer his "electromagnetic telephone."

In the end, he applied for a patent for his harmonic telegraph with the added claim that it could be used to transmit "vocal sounds." In other words, he applied for a patent for the telephone before he had a working model of one. And he did this on Valentine's Day 1876, exactly the same day that his rival, Elisha Gray, registered a patent caveat for the same thing. The resulting hoo-ha is now patent office legend, but suffice it to say that on March 7, 1876, Bell's patent was granted. Now all he had to do was get his transmitter and receiver right.

The very next day, his lab notebook bore a drawing of a "new" Idea, a "water transmitter." It looked remarkably similar to the one that Elisha Gray had included in his patent caveat. And it was by using this (inherited) water transmitter that Bell famously made the world's first telephone call three days later: "Mr. Watson, come here, I want to see you."

But on the day that Custer fought Sitting Bull, the water transmitter was nowhere to be seen. In the intervening months, Bell had returned to his designs on an electromagnetic transmitter. What he'd managed to do, with a string of intermediary prototypes, was tweak the resistance of his machine such that the voltage of the battery no longer blocked the transmission of sound. Bell, however, didn't know that this is what he'd done. All he knew was that the most recent of his many attempts had worked better than the least recent. He'd built lots of models, chosen those that worked best, and tweaked each of those winning designs in order to make the next generation of design. Over time, the models had gradually gotten better at being a telephone. He'd "bred" the telephone just as dog breeders in Britain bred the cocker spaniel.

In Bell's sweaty palm at the Centennial Exposition was his "best of breed." It looked a little like a funnel bolted on to a battery bolted on to a telegraph. With the equipment finally rigged, Bell asked an Exposition official to go to the receiver, half a mile away at the other end of Machinery Hall, and listen. As the audience chuckled, the fair official dutifully headed off across the second largest room on Earth and, finding a path between the bustled dresses and walking canes, disappeared into the crowd.

Bell craned his neck to look for the official, but he was lost. The hall was thick with visitors. Their incessant chattering filled the vast space with an echoic drone. But this human din was itself outdone by the slapping racket of water falling from jets and cascades coming from an area of the hall dedicated to hydropower. And this cacophony in turn was subsumed within the synchronized clanging of hundreds of frantic machines mercilessly torturing acres of windows at the command of a single 680-ton steam engine that puffed at the hall's center. Spurting water, ringing bells, chuffing pistons—each noise hailing the gradual construction of another Idea flip-book, lending its industry to the irrepressible hullabaloo of the late-nineteenth-century design world. It wasn't a great spot to try out a telephone.

Bell pacified those waiting about him with a weak smile and glanced pointlessly toward the other end of the hall. Having no cue to go by, he decided to turn and speak into the transmitter anyway. He repeated himself a few times, then, with nothing left to do, placed the transmitter down and waited with the crowd. Collectively peering at the anonymous hubbub, they were all eager for the first glimpse of the incoming official.

In fact, he was still hundreds of yards away, hurriedly squeezing between ladies in bonnets, doffing his cap automatically at any gent who caught his eye, bursting irrepressibly with laughter as he ran, racing back toward Bell. He spun along the fourteen-acre floor past the Brayton hydrocarbon engine, an ancestor of the internal combustion engine; the Line-Wolf ammonia compressor, a grandmother of the refrigerator; the Remington typographic machine, the first QWERTY typewriter; and the Wallace-Farmer electromagnetic dynamo, a precursor to the electric light. These exhibits were not the top attractions in the hall—like Bell's telephone, they were clumsy and a bit odd to the mind of the average fairgoer—but in retrospect, they were among the most important "creatures" in that room. Each was an unrecognized harbinger of the world to come, a world of private mechanized transport, of domestic machines, of mass media and piped power. Together with the telephone, the symbol of global communication, each lay at the very core of a future flourishing branch in the tree of Idea Life. They would each come to appear on some page or another in scores of future ancestry flip-books. They were all "Mitochondrial Eves." They had to come before what we have now, because Ideas can't just erupt out of nowhere. Ideas are like species: they must evolve.

A Space for Design

Those weird inventions became important retrospectively for one reason only: they opened up large, productive new regions of "design space." *Design space* is a term that has been used by R-and-D teams for some years. It is an imaginary, immeasurably vast volume in which all possible designs exist—all possible designs whether extinct, extant, as yet unrealized, or never to be realized. It's a tough place to envisage, but since living things are simply designs, design space will contain all the living things from every page of every ancestry flip-book in all of Life. All the unique designs of your direct ancestry, one tweak different from the last, going back 3.5 billion years, will be present in this space, together with all the unique designs of every other organism's ancestry. This mass of realized design is not present as one bubbling mob but is organized by one criterion: the complexity of design. In design space, the higher you climb the more complex the designs become.

In the case of animals and plants, we tend to call the species that appear higher up in design space "more advanced," but this often also implies "better," "more progressed." In reality, they are no better at living than any other living thing. Everything alive has succeeded in the task of continuing to live from the moment Life originated. The "more advanced" species are simply more *complex*. And they have become more complex not because they were destined to but because they were forced to when Life got hard.

Natural selection is not a constant. Some organisms are put under significantly more pressure to adapt by their selective environment than others. Off the Exuma Cays in the Bahamas, a sandy shelf just a few meters below the surface hosts what looks like a topiary garden. In fact, this is the world's only open ocean stand of stromatolites, rock colonies formed by cyanobacteria. Cyanobacteria are simple things. They are bacteria that can photosynthesize. Fossil cyanobacteria identical to the species that live off the Bahamas have been dug up that date back 2.8 billion years. For almost 3 billion years their design has not changed. That's because it was such a good one: a unicellular organism that can grow and reproduce quickly and fund its energy requirements with its own in-house solar panels. For nearly 2 billion years, the cyanobacteria had an easy Life; their stromatolites coated all the shallow, warm waters of the globe. They didn't need to change. But then, about 1 billion years ago, their abundance radically tumbled. Experts draw only one conclusion: an organism must have come along that was complex enough to eat cyanobacteria. This grazer would have arisen from a lineage that didn't have it so easy, from organisms living less inviting lives, on the fringes, in less inviting places. Its ancestors faced adversity, tough selective environments, and they overcame this adversity the only way that Life can: by adapting. Design work was done, and as it was done, the lineage accidentally accumulated complexity: mouth parts, digestive enzymes, a coelom; their hard Life forced them to get some guts.

Today, the stromatolites that once dominated the shallows are isolated oddities clinging on in the few spots that will permit them a simple Life. The rest of the world got complicated; Life got complicated. And the species with the most troubled pasts have become the most complex of the lot. Taking a winding path from fringe to fringe, they have accidentally accumulated design to such an extent that they now find themselves hanging high, high, high in design space.

The telephone Idea had a similarly demanding history. Human beings, as the crème de la crème of Dennettians, immensely value any idea that improves interhuman communication over large distances, and we give any Ideas that carry this promise a hard Life. Over the millennia, we've sent smoke signals, lit beacons, reflected sunlight off mirrors, waved flags, flashed gaslights—anything we could to send messages abroad. But our desperate desire for global communication and our perennial dissatisfaction with contemporary technology has meant that we're always hungry for the next killer communication app. When Michael Faraday discovered electromagnetic induction in 1831: first, new portions of design space opened up; second, natural selection whirred within the thousands of human minds that were aware of his discovery; third, these minds put all associated Ideas through the mill and forced them to adapt; fourth, these Ideas shuffled tweak by tweak into the new design space; and fifth, in less than a year, the first electromagnetic telegraph was switched on. Unlocking the telegraph unlocked the next region of design space, and the whole process started again. The telephone was now bound to be discovered, because it lay there, within reach, in adjacent unrealized design space. In one form or another, at some point, as some kind of accident of some kind of history; by some route through some minds, we were *destined* to get to the phone.

Bursting from the bowels of Machinery Hall on June 25, 1876, came the Exposition official, shouting. He *had* heard Bell. He *had* heard Bell's voice. The crowd around Bell fell into polite applause. Ladies tittered, and a few gents called, "Bravo," but their voices were instantly swallowed up by the commotion of the overexcited design space that panted like an organism all about them.

A Space for Genius

Far from all the fun of the fair, in the peace of rural Kent, England, on the morning of June 25, 1876, Charles Darwin was taking a solitary stroll in the grounds of his house. It was a Sunday, and Darwin's routine, ever since he'd stopped going to church in 1849, was to walk alone along the "sandwalk" while his wife and the rest of the village listened to the sermon and sang hymns. Indeed, Darwin would walk the sandwalk every day, completing several circuits. It was good exercise—he was now sixty-seven—but his real reason for going round and round the sandwalk was to find space to think. He called the sandwalk his "thinking path."

Uninterrupted thinking time was so important to him that he would place a pile of stones at the side of the path and kick one off onto the grass with each passing so that he didn't have to keep count of his circuits and ruin his concentration.

By that day in June, Darwin had had a lot to think about. For the first half of his adult life his mind had been consistently unsettled by the demands of his developing Idea. For the second half, he had been cursed with the mental labor of defending it. His Idea had consumed him. Since moving to Kent in 1842, he had become something of a recluse. His health had gotten steadily worse. He preferred his own company and, once the Idea was out there, generally refused any request to appear in public. He'd paid a high price for his Idea.

After his walk, at the same moment that Custer, half a world away, was preparing to go on the warpath, Darwin retired to his study to continue the task he had set for himself that summer: to draft "an account of the development of my mind." His objective was to explain in writing exactly how he had come upon his Idea, the theory of evolution by natural selection. But putting pen to paper, he must have soon come to wonder how he could hope to do such a thing. After all, where had the Idea come from?

The theory of evolution by natural selection undoubtedly originated in the primeval soup of other, more ancient Ideas on evolution that simmered away throughout Darwin's education and young adult life, the Ideas of people such as Lamarck and Darwin's own grandfather. But if Darwin was looking for an obvious ancestor to his Idea in the minds of others, he was unlikely to find it. The recent ancestry of his Idea had not lived out its life in the minds of his teachers or his colleagues, but entirely within his own mind, in that private territory I'll call Darwin Country.

In this respect, the evolution of Darwin's Idea was significantly different from the evolution of the telephone or the cowboy hat or the gambrel roof. Those Ideas were always present as a pool of many individual Ideas, resident simultaneously in many minds, shuffling collectively into the future as most species do. Darwin's Idea, on the other hand, had a host population of one—a situation that is ordinarily catastrophic for any Idea. Even worse, this Idea's lone host was a person who did not talk about the Idea to anyone; a host who, for years, just went on solitary walks, kicked stones, and scribbled indiscernible phrases in notebooks. But the Idea was lucky in at least one respect: it had found itself in

Charles Darwin's mind, and Charles Darwin was special. All on his own he would come to do the job that, for most other Ideas, it takes many minds to accomplish. On those walks, he singlehandedly hosted the evolution of an entire species.

If Ideas are like species, as these goggles suggest, then this is what was happening on those circuits of the sandwalk: Darwin was conducting an exercise in "Idea husbandry." He was captive-breeding the Idea, encouraging it to bud, cross-breed, and mutate, building up within his head an entire population of individuals of that Idea, each slightly different from the last. Then he was forcing this population of variants to perform their journey through Darwin Country, the selective environment of his mind, a new generation for each shuffle step. The variants that best suited the immediate geography of this unique landscape survived to father, within Darwin's continued musings, the next generation of the Idea. All the rest failed to survive. It was a brutal process—it was a hard life—but the harshness of Darwin Country forced adaptation, and enabled the Idea to continue its journey. And that's how it went for years: the Idea shuffling through Darwin Country, while its host clocked thousands of circuits of the sandwalk, adapting to the demands of his mind, tweak by tweak, generation by generation. But ultimately, as a result of all this mental industry, this intense design work, the Idea came to fit Darwin's mind environment, it reached the Promised Land, and Darwin's unique breed of the theory of evolution was ready to be released into the wider, wilder world.

Sitting at his desk, even as an episode of violence erupted in southeastern Montana, Darwin would not have recalled any of this detail. He would have remembered his head full of thoughts and the effort of concentration. He might have remembered moments of "epiphany," the infrequent occasions when a tweak turned out to be surprisingly fortuitous, or a mutation arose that expressed a useful trait. These were the moments when the Idea accidentally discovered a "new pass" or "river crossing" and forged ahead. And they may have been the moments when he grabbed his pen and scribbled in his notebooks. The scribbles mean little to us now—and possibly meant little to the older Darwin in 1876, sitting at his desk trying to rebuild history—but for the man walking the thinking path on a daily basis, they were desperate attempts to hold on to an Idea species in all its glorious variation at different stages in a "descent through modification." They were the best portraits he could sketch.

And hence those notebooks are the closest we could hope to get to an ancestry flip-book of the theory of evolution by natural selection. Bend them back, place your thumb on the corner, flip, and you'll witness the generations of that Idea fly by, the Idea as it adapted over the years within Darwin's mind. And that's what's so special about the task that Darwin was performing in the first half of his adult life: he was building a substantial portion of a species' ancestry flip-book and forging into new regions of design space all on his own.

We've got a name for people who perform such tasks. We call them geniuses, those loners who appear to summon spectacular, revolutionary Ideas "out of nowhere." But if these goggles are showing me the true picture, no Idea can come "out of nowhere"; it just feels like that because geniuses are special people with special minds, able to juggle whole generations of an Idea and test it in their meticulously constructed mind environments. When, after some time, the Idea emerges from this space into the wider noosphere, it's found to be hundreds or thousands of flip-book pages along; frequently to the great surprise of the genius himself, who—and he would be the first to tell you this—was conscious only of being a host to an unqualified process of "Creation":

> When I feel well and in good humor, or when I am taking a drive or walking after a good meal, or in the night when I cannot sleep, thoughts crowd into my mind as easily as you would wish. Whence and how do they come? I do not know and I have nothing to do with it. Those which please me I keep in my head and hum them; at least others have told me that I do so.
>
> *Wolfgang Amadeus Mozart*[1]

On the day that Bell gave his first public demonstration of the telephone and Custer died on a small hill in southeastern Montana, Darwin sat at his desk unaware of whence and how his Idea had come about. A month and a half later—on the day, in fact, that Wild Bill Hickok was shot in the back of the head—Darwin closed his autobiography with a paragraph of striking modesty, typical of a mind that has been sequestered to host the evolution of a great Idea:

> Therefore my success as a man of science, whatever this may have amounted to, has been determined, as far as I can judge, by complex and

diversified mental qualities and conditions. Of these, the most important have been—the love of science—unbounded patience in long reflecting over any subject—industry in observing and collecting facts—and a fair share of invention as well as of common sense. With such moderate abilities as I possess, it is truly surprising that I should have influenced to a considerable extent the belief of scientific men on some important points.

Darwin didn't have a brilliant Idea; the Idea had a brilliant Darwin. There *is* a space for human genius in this new world view. Indeed, through this new world view we come to see exactly what human genius is: it's the peculiar situation whereby a significant portion of design space is conquered by the evolution of an Idea in just one mind.

10

How the West Was Won III: America Making

The Maul of America

The small hill that I'm standing on in southeastern Montana is Last Stand Hill. It's silenced by the wind and the ghosts not only of the buffalo, but also of Custer and his men. In front of me is a fenced area that contains a liberal scattering of small, pale gravestones, one of which is painted black so that the name "G. A. Custer" stands out, even against the bright afternoon sun. Custer was standing on this spot more than a hundred and thirty years ago, also in the afternoon, wondering what to do. He could now see the size of Sitting Bull's village. The tepees spread for a mile along the valley below, just the other side of the dark green cottonwood trees that still stand on either side of the river. There were

2,000 Sioux, Cheyenne, and Arapaho warriors in those tepees—and up to 5,000 others. Custer's colleague Major Reno had already discovered the size of the village when they hit the southern end. Reno wasted little time in retreating recklessly back across the river, leaving Custer trapped, alone, with 248 cavalrymen. The Indians, led by Chief Crazy Horse, split the U.S. troops and picked off small lots of them, one at a time. Marooned on Last Stand Hill, Custer and 41 men shot their horses in a circle to use their bodies as barriers to the bullets and arrows. Shooting your horse is the last act of a cavalryman. They knew they weren't leaving this hill. The grass was thick—it was late June—and the Indians were able to creep up unseen. Custer and his men endured for less than half an hour. His Last Stand swiftly collapsed, and twenty-odd men finally threw down their weapons and ran for the ravine below the hill. The warriors pursued them on their painted horses, trotting alongside them, lancing them, and "counting coup." Later, Indians who had fought at the Battle of the Little Bighorn would liken this final act to a buffalo hunt.

An old National Parks ranger stands before us tourists retelling the story. He's full of passion; his face is animated. We're all hushed, because when you can see the theater of the battle before you, when you're only feet away from the dead, and when an old man is telling a story, the drama of the event pops into 3-D like a novelty helium balloon. "Custer was a great American," the ranger says, "and Sitting Bull was a great American. They both fought hard for the America they believed in."

Sitting Bull's victory at Little Bighorn knocked the wind out of the Grant administration, but the holy man knew that his glory would be short-lived. He understood that his outrageous triumph would not be allowed to continue. What happened on this hill was to mark the end of the Plains Indian way of life. The following year, Crazy Horse was killed and all but a few of Sitting Bull's followers moved back to a reservation now swarming with gold diggers. Sitting Bull fled to Canada, but after four years without regular buffalo, he and his family were starving and he agreed to surrender.

He lived on the Standing Rock Reservation in South Dakota and refused to sow crops. He made money by touring with Buffalo Bill's Wild West Show, the old buffalo hunter and the old Indian chief performing together for the crowds across the States, on occasions indulging in a dramatized reenactment of Custer's killing. In 1890, still causing strife, Sitting Bull was arrested by Sioux policemen. A struggle broke out, and he was shot in the head. At the moment he died, his gray horse, standing some distance away, rose up on its back legs, turned around, and started to dance. Many present thought the animal was possessed, perhaps by the departing spirit of Sitting Bull, but it was just an old circus horse doing what it was trained to do: rear up at the sound of gunshot. History rarely has such glamorous moments.

"The cultures clashed," says our old ranger, "I don't know why they had to, but they always seem to, even today."

I know why . . .

Roughly fifteen million years ago, the Pacific Plate of the Earth's crust collided with the Caribbean Plate, and began to slide underneath it. The massive friction that resulted lit a chain of underwater volcanoes in the shallow sea that separated North America from South America. The activity of the volcanoes crafted small islands in the sea, and the shallows around those islands trapped sediment. By three million years ago, the islands and sediment had linked to form a continuous land bridge between North and South America: the Isthmus of Panama. For the first time, large animals from each continent could cross to the other. Tapirs, bears, deer, pigs, camels, otters, raccoons, wolves, and big cats went south, while ground sloths, glyptodonts, terror birds, and marsupial carnivores went north. The event is known among biologists as the Great American Interchange, which makes it sound like it was convivial—just an evenhanded swap. It wasn't. The South American species fared far worse than the North American species. Not one species from the North became extinct as a direct result of the interchange, while hundreds of southern species disappeared within a few hundred thousand years of the making of Panama. The southern fauna just couldn't cope. A few species managed to establish populations in the North—most notably the armadillo, opossum, and porcupine—but they did so only because their peculiar approach to Life meant that they faced no opposition; instead of replacing parts of the northern menagerie, they added to it. In South America, it was a different story. There was an out-and-out invasion, a maul. All the sizable southern animals were replaced by the eagerly diversifying northerners. Hundreds of southern species vanished. Why was the interchange so one-sided? Because the history of the two continents was so different.

North America had, for a significant period of time, been connected to the rest of the world by Beringia, another land bridge, one that joins Alaska to Siberia. Granted, at many points in recent history the ice has melted, turning Beringia into the Bering Sea—as it is today—and barring any traffic of large animals, but for much of the last thirty million years, species have been able to wander freely between North America and the Old World. South America over the same time period was adrift, on its own. Hence the animals of North America were effectively derived from a stock of species that spanned an area six times that of the animals of South America.

In evolution, space is as important as time. The huge pool of habitats across the Old World and North America was responsible for creating a huge pool of species, and these species, unable to hide away on their supercontinent, had to compete with all their neighbors. The North American species had a much harder Life than those living on South America over the same time period, and this forced them to climb higher and higher in design space. They became more complex, with faster metabolisms, bigger brains, and greater efficiencies. They were the latest, highest designs on earth, the products of the biosphere's most competitive arena.

When Christopher Columbus accidentally discovered the Americas in the late fifteenth century, he played the same part as Panama. He connected the Old World with the New, and in doing so, he paved the way for a uniquely human "interchange." It was also a desperately uneven swap, a one-way flow characterized by three forms of invasion. The first was the invasion of Old World genes. The second was the invasion of Old World germs. The third was the invasion of Old World Ideas. Of the three, the last has had by far the most profound effect on the endemic American population. Yes, Métis were born. Yes, smallpox wiped out nearly all of the Mandan. But when you see Plains Indians today, in their blue jeans and cowboy hats, unable to speak their native languages or raise a tepee, you realize that it was the invasion of those Old World Ideas that had the biggest impact. And when you picture Ideas as species, you can start to see why that was. The species of Ideas from the Old World were simply at higher positions in design space than the endemic American Ideas. They had evolved in the noosphere's most competitive arena: Western civilization. They'd had a harder Life, competed with more rivals, moved through a larger pool of mind habitats. The rules of natural selection would suggest that these more complex species were destined to replace the endemics and that, because of the degree of the imbalance, they would do so at a breathtaking speed.

In the tidal wave of invading Ideas, a few endemic Idea species did manage to hang on—the powwow, the headdress, pemmican—but they did so only because their peculiar approach to Life meant that they faced no opposition. Instead of replacing parts of the Old World menagerie, they added to it. Where New World Ideas came into direct competition with the Old World species—the languages, the religions, the tepees— they just couldn't cope. It wasn't anything to do with which Ideas were "more advanced," "better," or "more progressed"; it was merely a question of how high they were in design space.

When two different cultures cohabit the same region of the noosphere, two different communities of Ideas find themselves competing for the same mind space. There will always be a bout of survival of the fittest, and where two species compete for exactly the same niche, the species that hangs highest in design space will prevail. That's bound to happen, because Ideas *are* like species. We become wrapped up in the maul between competing Ideas, fighting each other until we have to shoot our horses, only because we are their hosts.

Making America

So what about America? Was there enough time for the watch of infinite involvedness I spied from that window in JFK to have sprung up mindlessly? From the stories of a few years in one part of America—the buffalo hunts, the bank panics, the gold rushes, the Old World invasions—the history of this nation certainly appears to be mindless. But would mindlessness alone work fast enough?

To answer this we need look no farther than Deadwood. What happened in Deadwood Gulch between 1875 and 1880, under the astonished gaze of the Sioux, was a microcosm of what happened across America between 1492 and 1979 (when I was ten). We have to scale for time: each year in Deadwood represented a century in America; each month, just less than a decade.

In the summer of 1875, Deadwood Gulch was empty. But toward the end of that year, white folk arrived to mine gold. At first there were only a few, but in the months following, there was an alien invasion. The newcomers scrambled among themselves for the gulch's resources, and in doing so they permanently altered its environment. Their sweaty, soiled living conditions and high population densities soon seeded the second invasion: by mid-August 1876, smallpox was ravaging the town. The Sioux had little natural protection. This disease had had its own genesis among the world's first human cities in the Middle East thousands of years before. It was not a disease that the Sioux immune systems were prepared for.

But smallpox wasn't the end of it. Before the winter closed in, the third invasion was well under way: the invasion of Old World Ideas. By the close of 1876, civilization had sprung up in the gulch. There were saloons, hotels, retail establishments selling all sorts of strange wares, a regular stage coach service, gas lighting, and a telegraph. Just two years after that—two years after Bell's first public demonstration at the Centennial Exposition in Philadelphia—Deadwood had its first telephone. Later still came the railroad and electric lighting. The Sioux, watching the gulch fill month after month with yet more new Ideas, may have imagined that there was no end to the white man's invention, but this wasn't spontaneous generation; these Ideas had had their design histories elsewhere. The makers of Deadwood were simply capitalizing on a huge pool of Ideas that already existed over the eastern horizon. Or, to put it another way, the Ideas were capitalizing on new opportunities that had opened up in the West: new hosts, living in new places, new niches. Idea Life had gone prospecting.

And that's the story of the whole of America: an accidental discovery, an automatic human invasion, a mindless scramble for resources, and a subsequent, predestined regime change in the world of Ideas. There's nothing very glorious about it. As the Old World humans spilled onto and throughout the continent, fulfilling their own particular needs, their Old World Ideas had a free passage and, from their hosts, mounted their very own invasion and scramble, as mindless living things are prone to do. The speed of America's "rise" was so impressive only because the Ideas of the Old World and the New occupied such different heights within design space. When the colonies were formed, and a "mind bridge" erupted up through the Atlantic Ocean, the Old World Ideas were bound to conquer the new territory, like the big cats that trotted joyfully across Panama three million years ago to dine on the sitting ducks of the Amazon jungle.

A significant proportion of the making of America, therefore, was a mindless "correction," a flurry of Idea activity brought about by the accidental opening of new niches in the noosphere. O'Sullivan was right: because of the way culture works, the West was always destined to be "won." But that doesn't mean to say that all America making was at the left-hand end of the spectrum.

These goggles, it has now become clear, are not blind to human genius. Human genius, through these lenses, is, at its uttermost, a capacity for a single mind to host a private selective territory—an adaptive theater, a micronoosphere—that may, whether consciously (à la Darwin) or unconsciously (à la Mozart) bring about Idea evolution, generating significant additions to a species' flip-book on its own. This view concurs with Darwinian evolution. The essential component, a gradual descent by modification, is still present; the flip-books are still dull; it's just that their dullness is less evident because the genius is building the flip-book "behind closed doors."

And this realization finally rids me of the worries I first experienced while tipsy in Deadwood, my worries about Lamarckism. Individual Ideas go in to a mind, change their traits over several generations in order to adapt to the selective environment they discover inside that mind, then come out different from the Ideas that went in. This means that the adaptation that happens *within* a mind is the same as the adaptation that happens *between* minds. By adjusting my focus and seeing the Ideas for the minds, the ghost of Lamarck has been exorcised. It isn't a case of acquired inheritance at all; it's survival of the fittest, good old-fashioned Darwinism.

How do geniuses manage such magic? There's an obvious answer that I've missed until now. I've been so preoccupied with our status as the crème de la crème of Dennettian creatures (the ultimate thought swappers) that I've forgotten we are also supreme Popperian creatures (wondrous problem solvers), able to build exquisite models of the universe in our heads. What is the purpose of these models? They are places in which we test our Ideas before we implement them. That sounds exactly like a form of internalized natural selection, so my guess is that geniuses are able to take hold of the wheel and guide the world of human thought into new regions of design space simply by exercising their Popperian faculties.

This should give us all hope. We are not just Idea vehicles, conveyors of thoughts from one person to the next. We can, individually, make a considered difference. The capacity for genius is there within all of us; it's not the preserve of the few. And if you find it hard to build a flip-book all on your own, then take Socrates's advice and bat the Idea between yours and another's mind, or among the minds of a whole group. Genius can (and does) work in collectives, too.

Yet, abandoned on this ghostly hill by the Ohioan tour group I hijacked—who head off in their baseball caps, white T-shirts, and high, white flannel socks toward Deep Ravine to see the spot where the last of their fellow countrymen (the ones who fled from their dead horses in a blind panic) were butchered to a man—I'm not feeling particularly comforted by that notion. We may all have the power to participate mindfully in cultural evolution, but there isn't much in history to claim that we do so with any regularity or skill. We are far too passive in our role. Just as we would rather learn a solution than devise one, we too readily accept the culture we receive, and too rarely question it. This potted history displays how content we are to live our lives unquestioningly Although we appear to have our pilot's license, we tend to leave this plane of ours in autopilot, no matter how bumpy the flight. There have been too few America makers, and too few humanity makers. For the most part, cultural evolution remains mindless[+].

The Secret of Sitting Bull's Tepee

Ads and I drift over to the other side of the hill—the north. There, overlooking the Little Bighorn River as it snakes up to meet the Bighorn and ultimately the Yellowstone River forty miles away, is the Indian Memorial, a much more modern edifice than the memorial to Custer's men. Its brushed-stone walls list all the Native Americans known to have died that day, and carry sketches of the battle by some of the warriors who were there. There are dedications "to the brave" and "to the noble," to those who risked everything that day to defend the "old way" of the Plains Indians.

If I'd been able to ask Sitting Bull how old this "old way" was, I suspect his answer would have been abrupt and defiant. "The Sioux," he might have said, "have been living in tepees since the beginning of time." But had that been his answer, it would have been the holy man talking. Sitting Bull would no doubt have been aware that the Sioux were in fact recent arrivals on the Great Plains. The maul he was having with Custer

on that hill was to protect an "old way" that was relatively *new* to his people. Just a handful of generations before, the Sioux had been living in the
woodlands of Minnesota, the same woodlands the Scandinavians were
later to cut down. There, they must have lived a very different "old way."

They were forced to abandon their traditional home there, in advance
of the Scandinavians, by a domino effect that tipped each Native American tribe westward into their neighbor's territory upon the arrival of the
Europeans on the East Coast. Coastal clans became mountain tribes.
Mountain tribes became valley people. Valley people became woodlanders, and woodlanders headed out onto the prairies to become Plains
Indians.

The domino that hit the Sioux was the Chippewa (or Ojibway) tribe.
The Chippewa were the traditional enemy of the Sioux and had repeatedly fought with them for ownership of the woodlands around Lake
Superior. When French trappers started trading with the Chippewa, the
Chippewa warned them of their enemy, the "Nadouessioux"—"adders,"
poisonous snakes—approximating a French spelling that we've inherited
(in its shortened, adapted form). To help the Chippewa in their battles,
the French gave their trading partners firearms, and the time of the
woodland Sioux came to an end. Fleeing their earth-lodge villages, their
gardens, and their domed bark wigwam camps, Sitting Bull's great-
grandparents found themselves in an alien place: the tallgrass prairie.
Their only option was to form a new culture, a whole new Idea community. Their only method was to beg, borrow, and steal Ideas from the
Indians who already knew how to live in a land without trees, to requisi-
tion a new "old way."

So the Sioux tepee, the only tepee I've seen so far, is a relatively recent
species, a modern incarnation of an older Idea. But where did the Sioux
get their tepee Idea? Whom did they copy?

I scan the colorful names of the dead: Black Cloud, Bear With Horns,
Chased By Owls, Noisy Walking. About a third of them were Cheyenne
warriors. Then I remember what Melvin at the Journey Museum in
Rapid City told me: that the Cheyenne tepee was almost identical to that
of the Sioux. Surely this suggests that they share a common idea ancestor.

The history of the two tribes is complex. Cheyenne and Sioux may
have died together on June 25, 1876, but *The Encyclopaedia of Native
American Tribes* informs me that they were not traditional allies. Just as

the Chippewa tribe was the domino that hit the Sioux, the Sioux, falling westward, was the domino that hit the Cheyenne. At the start of the nineteenth century, the two tribes were constantly engaged in skirmishes, both desperate to secure a chunk of buffalo land. The Black Hills was a key strategic base, a source of timber for warmth, tools, and tepee poles. In 1800 the Black Hills was Cheyenne territory, but not long afterward the Sioux, perhaps benefiting from their swollen numbers, a legacy of their easier life in the East, took the Black Hills from the Cheyenne, and the Cheyenne were pushed farther west, into Wyoming.

The two old enemies made up their differences before the Battle of Little Bighorn, but they were not friends. Nevertheless, their tepees were almost identical. Why? When Native American tribes clashed, it was quite common for the victor to take the surviving women and children as captives. It seems likely that in at least one of their many skirmishes, the Sioux managed to capture, as slaves, some Cheyenne women. In Plains Indian culture, it was the women who hosted the tepee Idea. Tepees were made and erected by women, so in capturing Cheyenne women, the Sioux also, unwittingly, captured individuals of the Cheyenne tepee Idea species. It would have been only natural for the Sioux to borrow this Idea wholesale in order to succeed at their new way of life.

But why did the Sioux not then keep the Cheyenne tepee Idea? Why did their Idea of the tepee change to the extent that we can spot the difference? The Cheyenne tepee had "slimmer smoke flaps, with extensions added on the bottom edge." Or, conversely, the Sioux tepee had *broader* smoke flaps, *without* extensions on the bottom edge. What explains this diversification in less than fifty years?

Up to this point, I've focused on whether cultural species adapt as biological species do, gradually self-designing to fit their particular environments. Evolutionary biologists call this process microevolution, because it entails the tiny shuffle changes that we see in the successive pages of ancestry flip-books. It's the day-to-day activity of evolution, the process most studied by Darwin and so well explained by his natural selection. I'm now confident that the noosphere microevolves in the same way as the biosphere, whether we are mindfully involved or not. But this leaves a further question. Darwin was aware that the way in which natural selection worked out there, in the field, would bring about another, higher-level consequence. Quite by accident, in mindlessly enabling species to adapt to their environments, natural selection will also, under certain circumstances, mindlessly bring about the formation of *new* species. "Speciation," or "macroevolution," the forking of the growing tree of Life, is the maker of biodiversity. It is the process that has blessed us with a world so full of different designs on biological Life that we still don't know how many different forms there are.*

So how does macroevolution work in the noosphere? Do Ideas speciate, building forking branches and ultimately "trees," as they do in the biosphere? Does this mean that there is one original Idea lying at the base of the trunk of cultural life?

It's finally time for me to chart my cultural finch mob. I need more tepees! Craning my neck to peer over the somber walls of the Indian Memorial, I find some. On June 25, 1876, the enormous Native American village lay to the southeast of Last Stand Hill. Today it's to the northwest, a huge aggregation of tepees standing proud among Little Bighorn's fluff of cottonwoods. But they're not Sioux or Cheyenne; even from way up here I can see that. The Sioux tepee was squat and wide; these are tall, thin, and summited by a mad flurry of long, long lodgepoles. From this distance, they appear to be a perfect hourglass shape. These are Crow tepees. It's the annual Crow Fair, held in Crow Agency, Montana, "Tepee Capital of the World." I hope they're not "hostile," because we're going in.

---

*Not that we haven't tried to find out. One famous attempt cloaked a single species of Panamanian rain-forest tree in insecticide and then caught the rain shower of insect carcasses that fell from its fumigated canopy in a skirt of plastic sheeting that surrounded the trunk. Casting aside all but the plant-eating beetles, the researchers discovered that 682 different types lived their lives in the canopy of that tree, and they estimated, based on other research, that of these, some 140 types lived their lives *only* in the canopy of that one tree species. Knowing that there are some 50,000 different species of rain-forest tree on Earth, they estimated from this sample, that there could be as many as 7 million highly fussy beetle species living in the canopies of the world's rain forests. And given that beetles account for about a quarter of all known life forms—as J.B.S. Haldane is supposed to have said, the Creator must have had "an inordinate fondness for beetles"—that puts the figure for the total number of unknown life forms into the tens of millions, sky high above the catalogue of 1.4 million that biologists have named so far.

PART IV

Who's Driving?

11

A Beginner's Guide to
Tepee Taxonomy

Among the Crow

To get from Last Stand Hill to the village of gleaming white tepees below, we have to head back to I-90 and head north for one exit. We get off the freeway at Crow Agency, the settlement that operates as the capital of the Crow Tribe Reservation. Descending on the slipway, we fall in to a traffic jam of scrappy cars full to the brim with Native Americans: elderly, children, youths. There is a hum of activity. Weaving in and out of the cars are other families on foot. The local launderette has a line stretching out around the convenience store. The convenience store is managing its customers on a one-in, one-out basis. We barely move in the mass of cars. I catch the eyes of a number of Indian faces: those long proud noses, those hitched-up cheekbones, dark skin, and dark eyes. The buildings in this area lack the finesse that Ads and I've become used to in the white, Anglo-Saxon "ranch world" up above this valley. Plaster drops off the walls. Garbage lines the sidewalks. Stray dogs patrol. Young braves strut past us and stare. It's a little intimidating but, given how the West was won, entirely understandable.

The car shuffles on through the town. We blindly follow the Crow crowd, and before we know it, we've actually come out the other side, into fields that sit among the cottonwoods down by the river. And that is where the tepees are. Raising our eyes from the road, we see them all around us, in their hundreds. Lined up in makeshift streets, each with a car in front. Up close, the tepees are huge, and set against the greasy green of the cottonwoods and the azure of the afternoon sky, their canvases glow a bright cream. They show the asymmetrical cone typical of this kind of shelter, appearing to sit back on their haunches, content in the sun, but they seem more upright than the Sioux tepee I saw in the Journey Museum, and much taller. Of course, these are made of canvas—a significantly lighter and cheaper material than the traditional buffalo hide—so it's not surprising that they are bigger in volume than the museum specimen, but this species of tepee is definitely supposed to be *taller*. The height is accentuated by long lodgepoles that stick up into the sky above the nest as far as they reach down below it, so that they assume the hourglass silhouette I spotted from Last Stand Hill. The lodgepoles are clearly a matter of pride. Each tepee owner has stripped his poles so that they gleam like bone, and they wave for attention, many accentuated with red ribbons flapping from their tops or with their feathery dead pine foliage left intact, to catch the eye. The poles are loosely gathered in the nest, clambering over one another in their urgency to soar skyward. They look wildly, deliriously untidy. The Crow, I decide, like their poles long and extravagant and have a sense of humor; but looking around as we search for a spot to pitch our meager offering, their joie de vivre is not apparent: we still feel as though we're not welcome.

As it turns out, we are entirely wrong about this. After half an hour of searching for a clearing among the tepees (during which our nerves are further teased by the Crow Agency chief of police, who swears at us for driving the wrong way around this pop-up town), I eventually pluck up the courage to jump out, walk up to a tepee, and ask two Crow if they would mind becoming the Crow unfortunate enough to have us pitching next to them. The old man and lady are sitting on deck chairs in the private backyard of their pitch, hand-picking pieces of chicken from a pot that lies steaming beside them on a log.

"Hi, my name's Jon, and I'm here with my brother, Adam, and we're from England."

"Where?" says the old man, his neck extending toward me, his eyes crumpling into a squint.

"England, Britain," I stutter. For a moment the chicken in his hand is stationary, its barbeque sauce dripping onto the dry leaf litter below.

"You've come all the way from England to come to the Crow Fair?" he asks, coughing at the implausibility.

"Yes. We're looking for a pitch for our tent, and I wondered—"

"Sure, it would be our honor." His wife nods to confirm that it would. "Set up on the other side of our truck."

Sort It Out

Some weeks after my biology teacher, Mr. P. Veale, wrote his list of the seven characteristics of Life on the blackboard, he took his chalk and began to write a second list. His aim now was to explain how biologists, once they have identified the living things from the nonliving things with the assistance of the first list, go on to organize the living things into a catalogue of living things—or, to use the correct term, a taxonomy.

"King Philip came over from Greece smiling," he announced with nothing more than a few twitches of his long-drop moustache. The sentence was a mnemonic, to aid us in our task of remembering the different tiers of the system of classification devised by the Enlightenment Swede Carl Linnaeus. Linnaeus's idea was to group the many different types of living things into a hierarchy of bigger and bigger groups based upon shared physical characteristics. In order from biggest to smallest (with an example in parentheses to give you some idea of scale), the names of Linnaeus's tiers were: kingdom (animals), phylum (animals with spinal cords), class (animals with spinal cords and mammary glands), order (animals with spinal cords, mammary glands, and opposable thumbs), family (animals with spinal cords, mammary glands, opposable thumbs, no tails, and a dentition of $2/2, 1/1, 2/2, 3/3 = 32$), genus (animals with

spinal cords, mammary glands, opposable thumbs, no tails, and a dentition of 2/2, 1/1, 2/2, 3/3 = 32, and a cranial capacity of over 600cm³),
and finally, species (animals with spinal cords, mammary glands, opposable thumbs, no tails and a dentition of 2/2, 1/1, 2/2, 3/3 = 32, a cranial
capacity of over 600cm³ and . . . well, who look just like us).

Linnaeus was working in a time before the Idea of evolution had in-
fected European society. It's a testament to his approach that upon the
acceptance of evolution, his classification system was in fact more useful
than ever. Linnaeus had grouped organisms by their shared characteristics simply because it was the most convenient approach. Post-Darwin,
however, his catalogue assumed a new role: as a beginner's guide to the
tree of Life. Now every living thing in a group shared a common descent
with the others—they were on the same branch—and every boundary
between groups marked a split in descent, a fork in the branches. For example, the shared characteristics mentioned in parentheses refer to kingdom *Animalia*, phylum *Chordata*, class *Mammalia*, order *Primates*, family
Hominidae, genus *Homo*, and species *sapiens*—which is, of course, our
classification. Within that classification is the suggestion that, since the
group of living things with mammary glands is larger and more varied (a
class) than the group of living things with opposable thumbs (an order),
mammary glands evolved before opposable thumbs. Conversely, it suggests that every living thing with opposable thumbs will have mammary
glands. If this is not the case, then Linnaean logic would dictate that either we've got the groups in the wrong order or that opposable thumbs
must have evolved independently on two occasions.

Thus the Linnaean system is a road map to the branching paths of 3.5
billion years of biological Life, if you construct it correctly. The problem
is that there are many, many ways to peel this particular onion; as with
all hierarchies, the "group making" will inevitably be a subjective activity.
Hence taxonomists, the biologists tasked with this activity, are generally
a restless bunch, constantly fighting about where to draw the lines between groups.

"The only group, or 'taxon,'" said Mr. P. Veale, writing the word *taxon*
on the board and underlining it three times, with three accompanying
chalk squeaks, "that is, in any way, objective, is the one at the bottom: the
'species.'"

In the early 1940s, two biologists, the German Ernst Mayr and the

Ukrainian Theodosius Dobzhansky, attempted to tame the unwieldy and ambiguous Linnaean catalogue of living things with a standard "biological species concept." Their definition was that a species was "a group of organisms able to interbreed with each other to produce fertile offspring." In one sentence they gave taxonomy its first objective measure. If a mature male and a mature female organism can have babies, and those babies can themselves have babies, then the male and female organisms in question are indeed members of the same species. If not, they are not. Simple and precise.*

As well as giving a much-needed anchor to the taxonomists, the biological species concept was also the key to unlocking the secrets of what Darwin had called the "mystery of mysteries": the origin of species. Despite the title of his book, Darwin had never really explained how new species originate. He certainly had an answer for how species change over time, how they adapt, and with that he was justified in identifying a case for a single origin of (all) species. However, the details of how *new* species originate, of what happens during the act of speciation, were restricted to a guess that the finch mob he found in the Galapagos may have evolved *after* a migration event.

Mayr and Dobzhansky, working after the discovery of Mendel's Ideas on genetics, took over where Darwin left off. Their biological species concept could assist in explaining speciation. If a species is a group of interbreeding organisms, then a species is a group of organisms able to swap genes, and hence a species is a group of organisms with equal access to a "gene pool." But because of the definition that they themselves set, the species line has to be drawn at the very shores of this gene pool. So the correct way to conceive of a species is that it is a reproductively and genetically isolated group of organisms: a bunch of individuals trapped in a pool of their own genes, unable to paddle out of their pool or into anyone else's.

The making of new species was now easy to picture:

---

*Of course, we have to talk in general terms. Infertility is a physiological problem faced by all species. Just because your friends Bob and Hilda can't conceive a child, it doesn't mean that one of them is a member of another species.

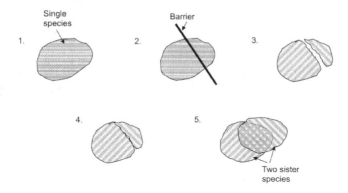

Speciation.

1. A species plays in its own gene pool, minding its own business.
2. Splash! A barrier comes down, dividing the gene pool into two, preventing "gene flow" from one side of the pool to the other.
3. The Force of nature does its normal thing: it mindlessly brings about change, but because the two parts of the gene pool are isolated from each other, there is no "averaging" of this change across the species, and the two part-populations are prone to change in different ways; they diverge.
4. After some time, this change is such that even if the barrier is removed, the two part-populations are not able to produce fertile offspring together.
5. Thus the two parts of the gene pool will forever remain separate, unable to patch up their differences, and two reproductively isolated "sister" species exist where there once was one.

Easy. The only thing to sort out was what was splitting the gene pools in step two? What were the barriers to gene flow?

The jury is still out on this one. We're absolutely certain that we understand one of the barriers, but there is continued debate about the viability of other candidates. The one we are certain about probably barricades gene flow in at least 90 percent of the cases of speciation; some say 100 percent. It's well demonstrated by the case of Darwin's finch mob, who were blown out to sea and marooned on a desert island. The first barrier to slap into the gene pool of the dull-colored grassquit was clearly six hundred miles of raging Pacific Ocean, a "geographical barrier," the easiest barrier to gene flow imaginable. If two part-populations can't get to each other to interbreed, then obviously there will be a speciation event in time. But the pool-splitting of Darwin's finches didn't stop there. Once safely on the Galapagos, their gene pool suffered at least thirteen further barrier incidents in a relatively short period of time. So what was going on?

The Galapagos is, you will remember, a double archipelago, a chain of oceanic islands with a chain of habitat islands on top. Birds as small and flitty as finches can conceivably be stranded on even these localized islands, so it's possible that geographical barriers were responsible for all the subsequent forks in the finch tree. But there are two other options:

⚭ ECOLOGICAL BARRIERS. Within any one living community, in any one habitat, there is a plethora of "ecological niches," occupations that a species could adopt, and hence adapt to. Remember the sharp-beaked ground finch? It turned its beak to everything from pecking at seeds, to snatching insects, to stabbing seabirds and sucking their blood. Depending on which sharp-beaked ground finch you see down the length of your binoculars, you could be looking at a granivore, insectivore, or parasite. These very different lifestyles may well require very different activity patterns. For example, perhaps the insect-eating sharp-beaked ground finch needs to be active at dawn and dusk, when the insects rise, while the vampire finch needs to work at midday, when the seabirds are resting. If so, the two part-pools of sharp-beaked ground finch might be present in exactly the same spot on the same island, and yet still never bump into each other. Naturally shift work like this would be a barrier to interbreeding.*

⚭ COURTSHIP BARRIERS. In 2007, researchers[1] discovered a new finch species in the making when they found that the medium ground finches on the Galapagos island of Santa Cruz now tend to be born with either a large beak or a small beak, but rarely a

---

*The husband of a friend of mine does night shifts, and my friend confirms that interbreeding when you are time-compromised in this way is indeed a genuine issue.

beak in between. There must be an ecological reason—the two part-populations must be specializing to crack open slightly different seeds—but the barrier to gene flow looks to be coming from a resulting behavioral accident. The two beaks appear to change the voice of the bird, and because a male bird's voice is its chief calling card, the two part-populations are serenading two different part-populations of females. Should this continue unabated, the gene pool will split, and another finch species will burst (gradually) into the world.

That's how species come about, said Mayr and Dobzhansky. Species are "reproductive isolates," so they must come into being upon "reproductive isolation" of one form or another. A split in a gene pool, brought about by a barrier to gene flow, is the ultimate cause of every fork in the tree of Life. Only upon the splitting of pool after pool after pool has Life become so diverse. Repeated isolations explain the whole forking tree.

Tongues in a Twist

Once our tent is up, Ads and I grab our essentials, wrap our tops around our waists—in the United Kingdom, the top-around-waist is dominant to the top-over-shoulder Idea; not so in Continental Europe and America, I notice, so it immediately makes us look different—and head off into the temporary tepee neighborhood that today stretches over several fields at Crow Agency.

Cast over the entire fair, via numerous loosely tied loudspeakers, is the urgent and bossy voice of a man speaking in Crow. We can tell by the high pitch and slight exhaustion that he is an old man, but he's not short on energy, incessantly bellowing updates on the proceedings and clearly taking the opportunity to mix in a few Ideas of "his own."

His words are not distinct to my ears, but I can feel the flow of his sentences. They rise and fall in a wail, with sharp summits and deep hollows, striking syllables in rapid succession. It's not like any language I've ever heard before, and its exotic color teases a smile from Ads and me simultaneously. We've landed in Indian country.

If I wanted to chart a taxonomy of culture, I could do worse than to begin with languages. Over the last few centuries a cohort of academics, the historical linguists, has in fact already done the work for me. Using the same approach as the taxonomists of nature, they have put in endless hours organizing the world's six thousand-plus languages into a hierarchy of relatedness. They've sought similarities among languages and grouped them into families of common descent. They've tied families together into a vast tree of language evolution with the implicit suggestion that language evolved only once in human history, and that all subsequent languages have thus derived from the original "mother tongue," separating again and again down through the ages. With a few notable exceptions, such as Latin and Sanskrit, the historical linguists have had to be content with working on existing languages; they don't have many language fossils in their collections. Not to be deterred, however, they have as one of their favorite engagements deducing the form of so-called proto-languages, ancestral languages, by using the features that related modern-day languages have in common to reconstruct their fossil parents. What do they find when they do this? That languages split in the same way and for the same reasons as biological species.

A language is an isolated pool of words. When a barrier comes down—splash!—into one of these pools, preventing "word flow" between two part-populations, a process of divergence accidentally/automatically begins. It happens gradually—so gradually, in fact, that none of the language speakers notice it's happening.

The first thing to change, because it's the easiest form of change, is the sound, the phonology, of the language. Vowel sounds are hit hardest because a vowel sound is pronounced with an open vocal tract—effectively a shaped mouth and throat and nothing else—and is therefore more susceptible to natural variation than a consonant, which is created by performing an exacting control on the pressure of air as it whips through to the outside world. One example of a vowel sound in the process of diverging can be found in my home country, England. If you ask a passing Englishman or woman to say the word *bath*, they will pronounce the *a* in one of three different ways. People from the Midlands and the North of England will say b*a*th, with a clipped *a* sound, which sounds harsh to everyone else. People from the South East will say b*ah*th, an open *a* sound, which sounds posh to everyone else. People from the South West and East Anglia will say b*aa*th, a lengthened *a* sound, which sounds hick to everyone else, or as an Australian friend once said, rather like a pirate. En masse, these sound variations within a single language, inherited and mindlessly selected down through the generations, form the "accents" that we recognize when people from different parts of a country come together. The accent is the first step in language change, and in modern times, because we can all chat so readily with one another, and because our languages are effectively dipped in preservatives by the habit of literacy, it's often the last. There isn't going to be a three-way split between the accents of England anytime soon. But if the barriers to conversation are maintained, the evolving languages will diverge further, until they become dialects.

Dialects differ from one another not only in their sounds but also in their vocabulary, grammar, and meanings. American English and British English are two large, umbrella dialects (with lots of other dialects contained therein). I once sat down with an American friend who was writing a prospective letter to companies in London. He started with "I write you in the hopes that . . ." No. No, no, no, no. Not a great start if you're trying to impress people with a British dialect. It looks all wrong, as if a five-year-old wrote it. "I am writing to you in the hope that . . ." is the way to start that letter in Britain. I told him so, and neither of us could quite believe that the other was so wrong. There is, of course, no right and wrong, even if the language is called "English," because languages evolve, and *evolution* and *progress* are not the same thing at all.

With time, and the further persistence of the barrier to communication, dialects will change to such a degree that they become mutually unintelligible. At that point, historical linguists will claim that the part-populations are now whole populations of new languages, and they'll adjust their taxonomic trees accordingly.

So, just as with biological Life, cultural Life appears to fork when information flow is prevented by some kind of barrier. Begoggled as I am, this fact could get me quite excited, but I can't be deaf to the linguists themselves, who caution that language evolution and biological evolution are not exactly the same thing. "The parallel with the evolution of animals and plants is obvious. The fit is far from perfect," says linguist John McWhorter,[2] an expert on creole languages.

Creoles are living, working, first languages that arose upon the merger of two or more parent languages. There are eighty-two listed on the Web site Ethnologue,[3] and the best anyone can do, because they defy taxonomic classification—they arise from *fusion* rather than *fission*—is to list them by the parent that is most dominant in the language. Thus there are thirty-one listed under "English-based," thirteen "Portuguese-based," eleven "French-based," and so on. Casting your eye down this list, you will be struck by one thing: Creoles tend to arise on islands or in jungles, and tend to be a combination of a local language and a European language. It's no mystery as to what happened: Creoles, in the main, arose from pidgins. A pidgin language is a "communication Band-Aid," a cobbled-together assembly of vocabulary and grammar, borrowed from any language at hand, in order that two or more mutually unintelligible groups of human beings can communicate. They're not fancy; they're just culture bridges that enable all parties to generate enough mutual meaning to get by. In recent times, pidgins have been scrambled together most frequently upon the arrival of Europeans in exotic foreign lands, either to manage the trade relationships that spark up or to enable plantation workers and slaves from various backgrounds to converse. They are articles of convenience and are dropped as quickly as they are picked up, but in certain circumstances they may remain in situ long enough that they become the dominant tongue in an area. In so doing, they are required to become richer and more decorated, so that people can convey more meaning more accurately with them. The speakers accidentally/automatically and collectively adopt new grammar forms and word morphologies to better deliver sense and tense. The pidgin undergoes a transformation from a lingo to a language, with the final step occurring when children are born who, exposed to no other talk, take up the pidgin as their mother tongue. From that moment on the pidgin is known as a "creole."

So the birth of a creole is an event exactly opposite to the birth of a species. Instead of a pool being split and diverging into two new forms, two or more pools merge, mix their information, and subsequently take on a new form. Little wonder, then, that McWhorter contests the analogy between language evolution and biological evolution. Because of the way in which creoles are made—and presumably have been made, just like this, over tens of millennia—the language family tree can never be a real branching tree, with fork after fork. It's more like the root system that you get beneath mushrooms in a woodland, a fungal "mycelium," a tangled net of interconnecting strands. Or, then again, as McWhorter says, like clouds, merging and drifting apart on the whims of the wind.

These goggles do suggest a response to this, but the time is not right. For now, I'm just going to point out that the tree or fungus or cloudscape of languages, whatever it is, is extremely useful to me in another sense. Languages are products of cultural history—their characteristics mark the journey of a people through the noosphere—and a more detailed knowledge of the Plains Indian languages will help me immeasurably in charting my cultural Galapagos.

Drummers in the Dark

The light drops quickly, as does the temperature. Ads and I untie our tops from around our waists, pop them on, and head toward the sound of drums. The Crow Fair is the largest in a season of annual powwows, festivals of drumming, dancing, and singing held at various reservations in the year in order to honor American Indian culture. I know that much, but I still don't have a clue what to expect as we join the Crow and pick a course around the tepees and trucks and toward the merriment. The sound of the drumming is, to this Englishman, like something from another age. At first you just hear the booming, speedy rhythm, but as you get closer you discern that there are voices dancing above the drum: semi-squealing, high, and sorrowful. My throat aches as I hear them. They draw us in like Sirens, but before we dash on the rocks, we come to our senses and discover that our path is blocked in the darkness by a thick aggregation of poorly parked, rusty trucks. Spotting our predicament, a young boy takes the initiative: "This way," he yells, and then spins off at a speed we can't hope to keep up with. Why do children insist on running everywhere?

Our scout appears back out of the darkness to find us, and beckons us between the bull bars to the shadowy backs of a wagon circle of fast-food vans. The air is thick with greasy globules, and the soil beneath our feet is sodden with melted ice. The boy encourages us to continue, and he hands our charge over to the thick scent of oily chicken chow mein that guides us through the last few shadows, past a huddle of diners on plastic chairs, and out into the main drag.

The natural light has gone now. We, and a thousand Native Americans, are lit with nothing but the reds, yellows, and pinks of the fast-food vans' neons. The Crow Fair is buzzing. Old men shake their heads and laugh with friends; young couples strut around, showing off their new babies; and all the teenagers are deliriously happy at the society of it all, skipping around in packs, their feet barely touching the muddy ground.

In the center, with its back turned to the circling throng, is a great shadow: a roofed central arena. This is where the drumming is coming from. We edge toward it, unable to see exactly where, or whether, we should go in. Then some people move within, and we glimpse the middle of the arena, bathed in light. A violent clash of colors shines under the lights. We see lots of movement but cannot pick out a single individual. We push through the audience, anonymous in the dimness, and the dancing competition is revealed: hundreds of Indians dressed in costumes designed to emphasize their celebration; costumes littered with bells, tassels, streamers, feathers, covered in paint; wearing colors that outdo their traditional dyes—ultra-gaudy pinks, greens, and yellows, highlighter pen colors—the colors a hundred generations of their ancestors would clearly have loved to get hold of. In this dance, getting noticed is the chief ambition.

The movement of the dancers is remarkable in its simplicity. They shuffle about a central pole—not a totem pole I notice, but a pole of lights and speakers—appearing to worship the music, paying penance by circling like monks about a sacred site, jogging in an endless repetition of small hops, their weight occasionally changing from one foot to another. The men dance violently, punching forward, crouching down, and jumping up. The women soar with their shawls, gliding from side to side, their moccasins barely leaving the ground with each hop. The whole spectacle is captivating, completely alien to Ads and me, who would judge a dance competition by drama and rhythm only. Here, the drama is in the costume, and the rhythm comes from something within these people; it doesn't fit the drumming at all.

As our eyes adjust, we discover that we are blocking the view of the people behind us, so we make a move for the seating stand that encircles the arena. When we sit down, we realize that sitting in front of us on small stools surrounding a huge drum is a circle of ten teenage Crow of all shapes and sizes, drinking soda. The boys laugh and chat among themselves. They occasionally use a stick to hit the drum out of boredom. They don't bother to watch the dancing, and I guess that they're just doing what teenagers do: finding a place away from their parents in which to mess about. The piped drumming comes to an end, yet the dancers continue to bob. Then, Ads and I get the biggest surprise of the trip: The ten teenagers suddenly raise their sticks in unison and begin to belt out a beat. They whack the drum skin so hard we jolt our heads back with each hit. After ten or so beats, one of the larger guys lets his head fall, creases his face up, and then lets out the most incredible wail, screaming out a series of open words so that the whole line sounds like the wind whistling through a contorted canyon. The rest reply with the same screaming, intricate song, peaking at pitches that must fold their voice boxes in two. Then all ten fall forward into a lower chant. I close my eyes and instantly feel that it's two hundred years ago.

When I open my eyes, I discover that a sound recordist is standing next to us, pointing a gun mic at the drum. It suddenly dawns on me that none of the drumming this evening has been piped: it's all been live, made by similar gatherings of drummers, hidden in the shadows all around the central ring.

A Pattern of Islands

This cultural Galapagos may have been devastated by the tsunami of Old World culture that washed over this continent in the wake of Columbus, but it's important that I become familiar with its once-upon-a-time geography if I'm going to be able to understand the evolution of the Idea species that have inhabited its shores. So here, again, is the basic scenario: before Europeans changed the world of the Plains Indians forever, this corner of the noosphere was a vast, sprawling archipelago of mind islands, thought-swapping communities that were, to a greater or lesser degree, isolated culturally from one another. Each and every tepee village on the plains was an individual island, a pool of thoughts discrete enough to sustain its own unique Idea adaptation. But none of these islands was entirely isolated. Villages had cultural relationships with one another—they met, they traded, they stole, they swapped suitors—so they were clustered in groups, of differing densities, just as real islands are in the South Pacific, brought close not by physical proximity but by cultural proximity: mutual intelligibility, mutual trust, and frequent contact. Lying between these clusters were varying expanses of cultural "ocean." On the noosphere globe, an ocean is any significant barrier to the movement of Ideas between minds. In the most barren parts of the Great Plains, this ocean was simply a span of peopleless, literally mindless, space. Elsewhere, tribes could be physically quite close yet culturally remote, divided from one another by substantial unintelligibility, mistrust, or the absence of contact. Idea species happily journey between tight clusters of islands, but large stretches of ocean present a challenge for even the most ambitious Ideas.

To draft a cultural map of the plains before the Europeans changed it forever, I therefore have to establish three things: First, which tribes were physical neighbors; second, which tribes were allies and enemies; and third, what languages they spoke. Of these three, the identity of the native tongue is perhaps the most important. Language, I would guess, was the chief determinant of the cultural geography on the plains in those times. Language is the medium by which most Ideas make their way between minds; language carries culture. So villages that shared a common dialect would have formed tight clusters; those with related dialects would have been located farther afield; and tribes that could not communicate at all, even if they were physical neighbors, would have had stretches of deep water between them. All of which would have radically affected the historic spread of the tepee Idea among the tribes. So finding out who could talk to whom is my first step.

Mutual intelligibility is only part of the story. Mere language relatedness is also important. Tribes with related languages may not have been able to talk to one another in recent times—after all, I don't know a word of Frisian, the closest related language to English—but their language relatedness is a sure sign that in not-so-recent times they could. Tribes with related languages have a common cultural heritage, and their Ideas, including their tepee Idea, should reflect this. Indeed, if that's not the case, I'm going to have to explain why!

Time to make a map . . .

1. PHYSICAL NEIGHBORS

By the time the first Europeans arrived on the plains in the early 1800s, up to thirty Native American tribes were using tepees. Of these, the majority were seminomadic, living most of their lives on the fringe of the Great Plains, in the woodlands, river valleys, and mountains, turning to their tepees only in buffalo season. Only twelve Plains Indian tribes were fully nomadic, living permanently on the plains, hunting buffalo 24/7. I'm going to start with the assumption that it was one of these twelve that first used the tepee, simply because, without a tepee, their lifestyle could not have existed. I may be wrong, but you gotta start somewhere. Where, then, did these twelve live?

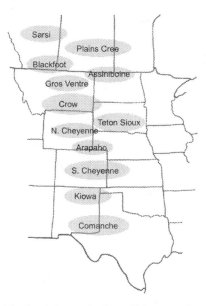

The chain of tribe islands existing on the Great Plains upon the arrival of the first Europeans.

In order to survive, each of the twelve nomadic tribes had to hold sway over a segment of the buffalo highway, the precious shortgrass that ran in a band two hundred to three hundred miles wide immediately in advance of the Rockies from Canada to Mexico. The north–south alignment of this highway was significant. Any habitat that runs north to south on Earth is ploughing through the climate zones. On the Great Plains the buffalo highway began, at the northern end, in near-tundra, and became, at the southern end, near-desert. Cultures are adapted to suit a particular climate zone: Ideas such as what time of year to settle for the winter, what timber to use for certain tasks, which plants can heal, and where to get water are specific to a climate zone. So it's not surprising to discover that, when the first Europeans arrived on the plains, they found the shortgrass divided up into latitudinal strips by furiously protective nomadic tribes. This is the order in which they were located, from north to south: Sarsi (in modern northern Alberta), Plains Cree,

Blackfoot, Gros Ventre, Assiniboine (in modern northeastern Montana), Crow, "Teton" Sioux (the westernmost of the Sioux, based around the Black Hills), northern Cheyenne, Arapaho, southern Cheyenne (in the center of modern Colorado), Kiowa, and Comanche (near the modern-day Mexico border).

2. ALLIES AND ENEMIES

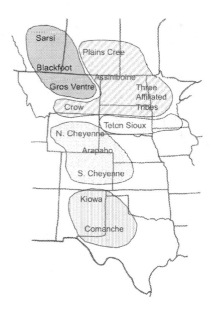

The five alliances of the Plains Indians upon the arrival of the first Europeans.

Gleaning what I can from my *Encyclopaedia*, I find that there appear to have been five different allied groups on the plains upon contact with the Europeans. A huge expanse of the northern plains was controlled by the powerful Blackfoot Confederacy—the Blackfoot, Gros Ventre, and Sarsi, who together appear to have been at war with everyone else. To their east and south was a second alliance, a close trading relationship between the Plains Cree, Assiniboine and, less intensely, with my new friends the Crow. This trade relationship was tied to the three affiliated tribes of Knife River, the seminomadic Hidatsa, Mandan, and Arikara, who, being river people, frequently brought in rare goods from the East and, being the closest settled people to the plains, operated a flourishing marketplace for the wider plains population. The Teton Sioux were universally despised, no doubt because they were new to the plains and fierce in their intent to force themselves on a crowded habitat, but they were of course closely allied with their woodland kin to the east. The fourth domain was the shortgrass in south Wyoming and Colorado, dominated by the allied northern Cheyenne, southern Cheyenne, and Arapaho (with whom the Sioux would eventually unite just before the Battle of Little Bighorn). The final alliance was between the Comanche and Kiowa, in the extreme southern plains. When French and Spanish expeditions first entered Texas and New Mexico in the early eighteenth century, they found these tribes at war with each other, and with their seminomadic desert neighbors, the Apache. By the start of the nineteenth century, however, all three of these tribes were in allegiance, perhaps united, as tribes in the North would later be, against their new foe: the European.

3. LANGUAGE FAMILIES

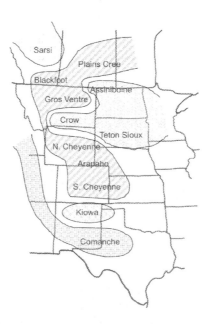

The language families of the Plains Indians.

The language associations on the plains thicken the plot even further by revealing more distant, unwritten cultural histories. The Cheyenne, Arapaho, Blackfoot, and Gros Ventre are all members of the "Plains Algic" language family, so at one point in the past they all spoke the same tongue. Arapaho and Gros Ventre in particular share many characteristics, suggesting that although their territories were separated by five hundred miles in 1800, the two tribes were, relatively recently, one ancestral tribe. It's possible to guess where this entire group of people originated, because they have a more distant relationship with the "Central Algic" language of the Plains Cree people, way up in northern Alberta and Saskatchewan, Canada.

The Plains Cree language is a member of a wonderful language phenomenon known as a dialect continuum. This is a row of mutually intelligible dialects lined up across the Earth (in this case east–west across southern Canada) that changes so markedly across its territory that if you picked speakers of the dialects from each end (in this case Plains Cree in the West and Innu over on the Labrador coast, in eastern Canada), they would not understand each other; they would be speaking different languages! Dialect continuums are examples of language change over not time but space. However, with time, should any member dialect break contact with its neighbor, new languages are inevitably (accidentally/automatically) formed. That's just how Life works. So a dialect continuum is a pool of information that stubbornly refuses to split.

Since the Central Algic continuum straddles the heartland of the whole Algic family, the cold woods and swamps of southeastern Canada, it's likely that the Cheyenne, Arapaho, Blackfoot, and Gros Ventre ultimately stem from there, too.

The Comanche, the Kiowa, and the Sarsi languages are all isolates on the plains. Comanche is closely related to the Shoshone language and, more distantly, to the Ute language, both of which are spoken by people in the central Rockies and Great Basin. So the Comanche people must have, at some point in their history, made their way from this region all the way down the front of the Rockies to the territory they held and defended so effectively in the early 1800s that it became known as Comancheria; this encompassed much of modern-day Texas, Oklahoma, and Kansas. Their northern neighbors, the Kiowa, are on a limb of their own, not obviously related to any large language family. The Sarsi, in contrast, are part of a huge language family, the Na-Dene group, which spans a territory from the Sarsi position, on the most northerly plains, right across northwestern America through the boreal forests and tundra to the tip of Alaska. Hence, despite being allied in recent times to the Blackfoot, the Sarsi probably stem from a very different culture that was adapted to life in the frozen lands to their north.

The Sioux, as Sitting Bull knew well enough, originate from an ancient culture that had its homeland in the eastern woodlands and valleys, and since the Assiniboine also speak a "central Siouan" language, the assumption is that they originate from that same ancient culture.

But what of the language of the tribe that Ads and I are now in the thick of: Crow? Even today, with highly contagious English buzzing constantly about the Crow's ears, the Crow language boasts more than four thousand hosts. The Sioux used to call it the "old tongue," and in a way they were right. The historical linguists tell us that Crow is a Siouan language, but it is not directly related to the Sioux spoken by Sitting Bull's tribe. It's more closely related to the language of the Hidatsa, one of the three tribes that lived high up on the Missouri in their earth lodges. Hidatsa is not the ancestral tongue (because Hidatsa is a modern language with a flip-book as large as Crow), but the two languages have very similar grammar rules, word forms, and sound forms. They haven't changed that much, which means that they were as one not so long ago. The degree of change, together with evidence from the descendent tribes' own oral histories, and the reports of the first Europeans on the scene, the French and British trappers, suggest that they were one ancestral tribe speaking one ancestral language as recently as three hundred to eight hundred years ago. Here's how I digest this news: the primordial Hidatsa succeeded in colonizing the Upper Missouri and began to flourish. As their population grew, the villages became too large for the resources of their hinterland and the *ur*-tribe was forced to propagate other villages farther upriver. Perhaps the inhabitants of these villages struggled more than those downriver because the environment was just that bit harder to tame. Perhaps they had to rely more on the buffalo hunt, and its pemmican, than the others. Perhaps they had to live for longer periods on the plains. For whatever reason, they became more and more dependent on the buffalo, until they reached a point when they remained on the plains even through the hard winter. At that point they became fully nomadic, and while the tribe they left behind became Hidatsa, they became Crow.

What tepee did they build to harbor themselves?

Poles Apart

Ads and I stand in front of a whole row of them, squinting into the morning sun. I can see that they are tall and white, with flamboyant pole tops, but what are the detailed features of this tepee that make it particularly Crow? While we stand gazing, a guy with a rounder face than most comes over to say hi. His name is Kenny, and he's from the Ute tribe, the tribe that gave its name to Utah and one of the seminomadic tribes that circled the plains.

"You can tell it's Crow by the smoke flap poles. See how they poke through a hole in the smoke flaps, and they've got a little stick tied to them, to stop the pole going all the way through; that's a Crow thing."

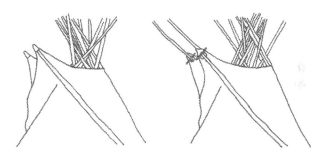

Tepee with smoke flap pockets.   Tepee with smoke flap holes.

He's right. Whereas the Sioux (and presumably Cheyenne) tepees have pockets on the corners of their smoke flaps to receive the smoke flap poles, these tepees, all of these tepees, have holes. The smoke flap poles can be as long as they like, because they poke right through to join the rest of the zany Crow poles in their race skyward.

"And of course the Crow use a four-pole tepee," says Kenny.

"What do you mean?" I respond, knowing full well that it took more than four poles to make the tepees in front of us.

"When you build a tepee," Kenny explains, "you start with a structure of either three or four poles, which you tie and hoist up into the air. You use that as a base to lean all the other poles on, and when you're finished, you put the cover on top. The Crow start with four poles."

This is a startling revelation, a fundamental split in tepee construction, an either/or situation—exactly the sort of thing to split a taxonomic tree in two.

"What about the Ute? What do you use?" I ask.

"We use a three-pole. Most tribes do."

I suddenly remember noting down that the Sioux tepee in Rapid City was based upon a tripod of poles. The Sioux were three-pole people.

"Why?" I inquire.

"It's easier. Well, *we* think it is. I can raise my three-pole tepee on my own. You can't raise one of these alone; it takes two of you," says Kenny.

I glance at Ads and smile. He looks at me out of the corners of his eyes as though he's trying to tell me that I may be getting too excited about all of this.

"Is it just the Crow who use four poles?" I ask, and begin to feel that I should perhaps explain the twenty questions.

"The Blackfoot do too, and the Sarsi, I think."

"So why do they use a four-pole if it's harder," I continue, hardly letting the man breathe.

"It's just the way they've always done it. They find it easier to do it their way, I guess."

"And which came first?" I say, before I realize that I've now gone too far. Kenny looks at me quizzically. I make another attempt: "How long have these two kinds of tepee been here?"

He stares, and then he looks away. "Since the Creation," he says, as if I'm talking nuts.

12

Bound by Imagination

The World Turned Upside Down

"Which are more advanced, three-pole or four-pole tepees?" That's the question I'm asking myself as Ads takes the wheel and points the Chrysler south again, so that we backtrack into Wyoming but then head off the freeway at Ranchester, and up into the Bighorns, the mountains that wooed us so much a few days ago.

If Kenny is correct, and the three-pole can be erected by one person but a four-pole takes two, then sense would dictate that the four-pole tepee is the predecessor. You can imagine four-pole people adopting the three-pole tepee, but not the reverse. To me, knowing virtually nothing about this as yet, the three-pole tepee, with its innovative tripod, feels farther outward and upward in design space than the four-pole.

But in contemplating this point, I find that something else is bothering me. As well as being a four-pole tepee, the Crow tepee is distinguished by the holes through which the smoke flap poles poke. This makes smoke flap holes a second distinguishing trait. The Sioux tepee was a three-pole tepee *without* smoke flap holes. Fair enough; that would support the theory that the three-pole/four-pole split was the fundamental split in the tepee Idea taxon. But what if, I ask myself as we begin to climb up toward a pine forest that looms above us, I find that the tepees of two other tribes are as follows: tepee one, a *three*-pole tepee *with* smoke flap holes; tepee two, a *four*-pole tepee *without* smoke flap holes. If these four variations of those two traits exist, then how am I going to explain that? What, in this fantasy scenario, is the order of descent?

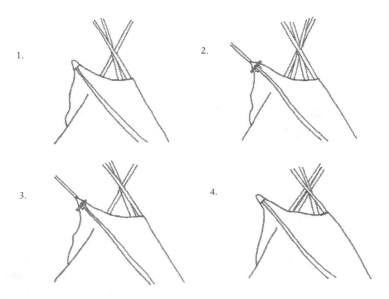

Four variations of two traits: 1. a three-pole tepee without smoke flap holes (Sioux); 2. a four-pole tepee with smoke flap holes (Crow); 3. a three-pole tepee with smoke flap holes (?); 4. a four-pole tepee without smoke flap holes (?).

If I stuck to Linnaean logic, I'd have two options: I could suggest that the tripod trait evolved before the smoke flap hole; but then that would imply that the smoke flap hole evolved on more than one separate occasion. Or, I could suggest that the smoke flap hole trait evolved before the tripod; but then that would imply that the tripod evolved on more than one separate occasion. Either way, the conclusion is that exactly the same trait evolved more than once: a remarkable coincidence, a reinvention of the wheel. Is it likely?

It's certainly possible. The biosphere is full of nature's coincidences. Evolutionary biologists call them examples of "convergent evolution," occasions in which organisms from separate backgrounds end up sharing similar traits because they adapt to fit similar environments. There's the case of the marsupial wolf from Tasmania that had a skull almost identical in every way to the skull of the placental wolf that we all know and love. There's the case of the treacherously armored plants of Madagascar's spiny forest, which appear to have copied wholesale the design plan of America's cacti. Then there's the complex eye that appears in the heads of both octopi and me. Coincidences do happen, but my problem is that convergent evolution isn't the simplest explanation in the case of cultural species, is it? The simplest explanation in this case is that the Plains Indian tribes were exposed to both the tripod and the smoke flap hole traits on separate occasions and either accidentally/automatically or intentionally/consciously combined the two traits, merging them within their one existing tepee. And if that was the case, then tepees are less like species and more like languages, clouds that drift apart and merge on an ad hoc basis. Nothing like octopi.

Or, maybe there's another suggestion, no less upsetting. What if the smoke flap hole and the tripod base are not traits at all but independent living Ideas in their own right? I've merrily referred to the tepee as an Idea species, and each memory of a tepee in an Indian's head as an individual member of that species, a cultural "organism"—but what if my scale is out? What if, in a tepee, I'm not looking at one organism but a whole conglomerate of organisms—a tripod Idea, a smoke flap Idea, a lining Idea, a doorway Idea, and so on—a rough assembly of symbionts living together as one "superorganism," giving the illusion of an independent thing in the noosphere? If that is the case, then I have no adaptive radiation to chart. The evolution of tepees is then a cultural epiphenomenon, an imaginary descent cast by a changing throng of smaller Ideas. So I've hit another problem. I've not yet gotten to my fourth tepee (or the top of this mountain) and I'm already questioning my interpretations of the words *organism* and/or *species* in the context of the noosphere. I'm already doubting the foundation of my analogy and the foundation of my comprehension. There's condensation in my goggles once again.

Ads picks up on my frustration, gives up on the drive setting on the automatic, and punches the stick into second gear. The Chrysler growls in complaint as it creeps up the switchbacks that bravely ascend the formidable eastern escarpment of the Bighorn Mountains. The going's uncomfortable, and when we see a "scenic outlook" up ahead, we gladly escape the road for some respite. Stretching our legs, we're met with the sight of an ancient geological catastrophe: sedimentary rock striped like a stockbroker's shirt in purples, off-whites, and oranges, buckled and contorted into bows and loops by unimaginable forces. The information board there tells us that the Bighorns were created in the Laramide mountain-building event that crafted much of the Rockies 70 million years ago. It was a "thick-skinned deformation," hauling up not only the surface rocks, such as the ones we see folded like modeling clay in front of us, but also the deep-seated, most ancient rocks, the 0.5- to 4.5-billion-year-old Precambrian shield that lies somewhere below all of us. Here, in this part of Wyoming, it is above us. The Laramide event flipped this part of Wyoming like a pancake: the young rocks are now below, the beginning of time above. Halfway up, we've already traveled 300 million years into the past, a time before the dinosaurs, and we'll continue to punish this Chrysler until we run out of time, the mountains' plateau, and we drive 8,500 feet up in the air on rocks as old as Life.

Seeing the world turned upside down like this has given me an idea. I think I've worked out a way of getting out of my predicament. It's a fairly drastic and unsophisticated response to the threat that my two made-up tepees have laid upon my analogy, but I can't think of an alternative. I'm going to repair my faith in the noosphere by smashing up your faith in the biosphere. The two mixed-up tepees I've concocted cause problems for me only so long as I allow you to keep alive your idealized concepts of the words *organism* and *species*; and, I'm sorry, but I can't let you do that. It's time for *your* world to be turned upside down. Let's start with the word *organism* . . .

At the three-hundred-million-year mark, the organisms buried in the strata about us are sharks, reptiles, insects, amphibians, and prehistoric trees—organisms that make sense: discrete, sizeable, independent living things that survived and reproduced and handed on their traits. Under their skins, running their target-driven lives, were selections of dedicated organ systems, functioning together like a machine. Each of those organ systems in turn comprised a series of organs that shared the load of bodily performance. And each of those organs was a congress of tissues that, upon some collective vote, carried out the tasks at hand. And each of those tissues was in fact an aggregation of cells that en masse lent their weight to a shared goal.

They were complex, exquisitely engineered organisms that sat perfectly where they were supposed to in Mr. P. Veale's third most important ranking system: the "levels of organization" rank he sketched out horizontally on the blackboard one day in June, ignoring the sneezes of a class of children at the mercy of a flowering willow and an open window:

genes—cells—tissues—organs—organisms—populations—communities—ecosystems

"Those to the left exist within those to the right," he announced, "and contribute to their function."

But it takes only another thousand feet and 200 million years for Ads and me to be among organisms that refuse to honor Mr. P. Veale's levels of organization rank. Jellyfish and algae, which are pretty much all that there is of any size at this point, were, and still *are*, no more than free living bundles of tissues. They float in the oceans or hold fast to rocks as mere collectives of cell types in cahoots, without organs or organ systems.

And another thousand feet above them, with the brow in sight, we crawl up into the youngest layers of a solidified Precambrian world, the Proterozoic, a two-billion-year span, where the life-forms are smaller, simpler, softer again. In these rocks the word *organism* refers to less than even tissues: sponges and extinct oddities not inclined to fossilize, and evident only from whispers of their forms trapped in billion-year-old mud. These organisms were loosely cooperating cells, nothing more—the first ever experiments in multicellular Life.

When we finally reach the lost world at the top of the escarpment, we've run out of Life-time. The dust that envelops our car, as the wind picks up on the plateau, comes from the Archean, an age before multicellular Life. In that world, two and a half billion years ago, every organism in the Archean was one cell big. Some grouped together to form colonies, such as the cyanobacteria I've mentioned, but they were colonies not organisms—together for their mutual benefit, but able to live alone if necessary.

So in the time it took to lay down this mountain, Life has borne organisms of all sizes and complexities as it has risen in design space. Depending upon the age you are in, an organism can be like you or me, or like parts of you or me. And conversely, you or I could be considered an organism or conglomerates of organisms.

The Medicine Wheel

Just as we are about to drop over the other side of the Bighorns and into the present, we take a break and follow a track farther up the mountains to a small hut and a long-drop toilet. It's distinctly chilly on the top of the world, so we wrap up in coats for the first time since our arrival in the States and head off, alone, on a path that wanders up into the hills. We soon rise above the tree line, and the pine forest begins to take on the appearance of a top-of-the-range shag pile carpet. As we clatter over the scree, marmots and pikas run for cover. The hill that flanks us to the west drops away, and we find ourselves on a ridge with the whole Bighorn Basin laid out to the west, thousands of feet below us, its parched landscape stark yellow against the reckless green of the mountains.

I start to feel the labor of walking. The air is thin as we approach ten thousand feet. We push onward, up a last incline to the meadows on top. Our heads begin to ache, but then ease as we see what we've been promised: lying prostrate and sprawling over some of the oldest land on Earth, a giant wheel of stones some eighty feet in diameter, with dozens of stone spokes and an unkempt cairn at its center, the Medicine Wheel, a sacred Native American landmark that for centuries has received the prayers and atoned the sins of Native Americans from scores of tribes. The signs are that it is still doing so: the structure is shielded from the casual observer by a simple rope fence, and along its entire length, this rope is draped in clutches of feathers and flowers, ribbons and flags, even seashells and animal-skin drums. There's a huge buffalo skull in the middle.

I can feel the vibe—the "spirits," I suppose. Just because I don't believe in skyhooks, it doesn't mean I'm numb to the aura of a place upon which centuries of human beings have focused their devotion. And the location doesn't hurt: here, where you can see the whole world, where you are cold and lonely, stung by the odd spike of iced rain, your eyeballs chilled and slapped by the zephyr that guards the mountaintop. But as my brother appears in view at the opposite side of the wheel, I can see that this particular sight-see is having a more profound effect on him. He wants to stay here for a while, and he wants to stay silent. When a group of jolly Texans arrive, I patiently wait for them to complete their circuit and leave, determined that we should have this place again for ourselves. When I look up, Ads has disappeared. I guess that he's gone to seek some spiritual solace somewhere nearby, so I resist the urge to find him, and instead sit down on a rock. I stare at the wheel, its peculiar bunting of of ferings flapping in the urgent air—and then suddenly, "out of nowhere," I get my own moment of clarity.

The huge monument in front of me may be called a wheel, because it looks exactly like a wheel, but it has nothing to do with the other wheel Idea, the one behind those circular rotating devices that orchestrate the mobility of vehicles and the operation of machines. I know this because *this* wheel dates from before the arrival of Columbus, and before the arrival of Columbus, there were no examples of the *other* wheel Idea in the New World. In place of a wheeled vehicle, the Plains Indians used a "travois," a triangle of poles that simply dragged along the ground, hauling their tepees and other worldly goods across the shortgrass under huge duress, carving tracks on the grasslands that are still evident today. So this medicine "wheel" is an example of convergent evolution in the noosphere, an object with undeniably similar traits but resulting from a completely different lineage of Ideas.

Sadly, the Ideas behind this medicine wheel are now extinct. No one alive today, not even contemporary tribal elders, remembers them. That said, the temptation to "reverse-engineer" the Ideas behind the medicine wheel is irresistible, and many have tried. They claim that the circle around the wheel denotes the Sun, or the Moon, or the Earth, or the cycle of life from cradle to grave, or home: the floor of a tepee. They say that the wheel is a clock, with twenty-eight spokes to reflect the human menstrual cycle, or the lunar cycle, or to point at the various sunrises and sunsets that open and close the four seasons. But sitting here, I realize that we will never know if these imposed Ideas are correct. We will never know the true meaning of this wheel or of the scores of other medicine wheels that lie on their backs on high spots adjacent to the plains, up and down North America. Now that the minds that contained those original Ideas have themselves turned into spirits, the best we latecomers can hope to do is sense the indisputable human and natural drama with which this type of monument is imbued, and bring to these remote locations our own desires for their meaning. And, in realizing this, I find my own meaning . . .

The wheel Idea with which we *are* familiar arose only once,* probably in Mesopotamia or the Northern Caucasus or Central Europe, and possibly around the middle of the fourth millennium BC. It's hard to know exactly because, upon its "invention," it was greeted with a *Dora the Explorer*–scale fervor, and the wheel rolled out at lightning speed across all the connected cultures of the Old World. Hardly surprising when you consider that some people had been tending farms for five thousand years without a wheeled cart. When the wheel came along, everyone got the Idea pretty quickly.

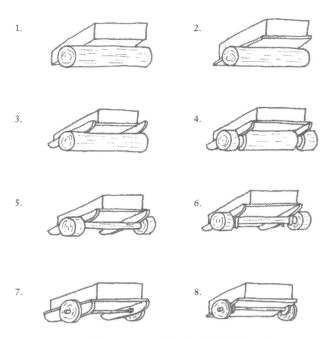

Some of the portraits in the wheel Idea flip-book.

*No one ever reinvented the wheel, you'll be relieved to hear.

But, once again, the question I'm forced to ask is this: Why did it take us so long to get the Idea in the first place? And, once again, the answer I'm forced to give is this: because Ideas evolve accidentally/automatically like living things, and that takes a long time. The wheel had a dull ancestry flip-book like all the other species of the noosphere. Here are some of the portraits that you might spot on a casual flip (from least to most recent):

1. A heavy load lies on a series of tree trunk "rollers." The load moves as rollers are repeatedly removed from the rear and added to the front edge.
2. A smooth plank board lies between the load and the rollers, reducing friction during movement.
3. The plank board now has blades underneath, so that it looks like a sledge balanced on top of the rollers. The blades reduce friction further by minimizing the surface area between load and rollers.
4. There are grooves in the tree trunks within which the blades rest. (The vehicle in stage three carves its own grooves in the trunks with repeated use, but because these stabilize the load, eventually they are added by hand);
5. The tree trunks are hollowed between the grooves so that different-sized sledges can fit on top (and an accidental axle and wheel pair is formed).
6. Pegs are attached to the bottom of the blades, in front of and behind the position of the axle, so that the axle is contained underneath the load and there is no longer the need to repeatedly place rollers at the front (that must have been a great day!).
7. The strength of the device is improved by the appearance of a separate axle and wheels, so that the axle can be fed through holes that have been fashioned in the blades of the sledge.
8. Ta-dah!

This process will have taken thousands of years and scores of human generations. There may well have been the odd genius involved along the way, conducting his or her own mind experiment, advancing the flip-book by a few hundred pages, but in the main, the invention of the wheel was a routinely get-rich-slow affair.

But why, why, why, WHY? What is it about the way that we very, very smart organisms engage with Idea evolution that requires us to take such tiny little steps? Sitting on a rock staring at a medicine wheel has made me realize that I already know the answer to this question: it is because the whole procedure of Idea evolution is cripplingly bound by our lack of *imagination.*

Imagineering

Imagination is such a wishy-washy "folk" word that psychologists try to avoid using it, but if pressed, most will admit that there is a faculty in our minds that enables us to hold memories together artificially in order to "picture" something that has not yet been. We must, in fact, possess an imagination, because without one, we couldn't have a Popperian intelligence. Popperian intelligence is all about problem solving, and problem solving requires picturing things that have not yet been. It's a very useful tool, and we love using it—research suggests that we devote an extravagant amount of time and brain space to thinking about things that have not yet been. Some researchers believe that our capacity to imagine the future is the defining feature of our species.[1] Unfortunately, our imaginations, despite being such fun to use, and so often used, are severely restricted. Let me demonstrate.

Imagine a blue sheep. It's easily done. You've seen a sheep before, and you know the color blue, so you have a perceptual memory of both entities, and because of that, it's easy to combine the two even if you have never perceived the two combined in reality.* That's your imagination working. After imagining a blue sheep, you could conceivably go out and dye a sheep blue, and the result would be very like the vision you've imagined.

Now imagine a coquelicot fosa.† Harder to do. Again, it's an animal painted a certain color, but only a minority of people will have a perceptual memory of both the animal and the color. If you are a French adventurer, you have a good chance. *Coquelicot* is an informal name in France

---

*I've actually seen purple sheep. In the north of England they have a habit, for some reason, of dyeing Herdwick sheep purple when they display them at country fairs, so it's an even shorter step for me.
†I know this is a good example, because both words are underlined in red in MS Word.

for a type of poppy, and the color *coquelicot* is poppy-colored: bright red with a bit of orange. The fosa is the largest carnivorous mammal on the island of Madagascar (once a French colony and partly French-speaking, so a likely destination for a French adventurer). It looks a bit like a lean and mean medium-size jungle cat, but that's just an illusion; it's actually an overgrown mongoose that, marooned on an island full of tasty lemurs, has been forced down the convergent evolution route.

Now that you have DIY memories of *these* two entities, you may be just about able to hotwire a patchy vision in your imagination. It's much more difficult than imagining a blue sheep, because, unless you *are* a French adventurer, you've never (knowingly) seen these two things, so in imagining a coquelicot fosa, you're combining two memories that are themselves imagined combinations of genuine memories—the coquelicot fosa in your mind is an abstraction, a vision one step removed from memories of things you've actually seen.

What if I asked you now to imagine an alizarin fanaloka? And then I told you that alizarin was like a purplish coquelicot, and a fanaloka was a smaller, furrier, stripier fosa? It's starting to hurt now, isn't it? Even with all those useful tips, imagining that extra step is too much. You can't fix the image in your mind because you're too far removed from genuine memories, from perceptual inputs. The image of the alizarin fanaloka struggling to appear would be an abstraction *of* an abstraction—two steps removed from your perceptual memory; too far removed from your experiences to be easily imagined.

It was the same with the people who brought us the wheel. They could only ever work with the memories they had, the firsthand memories of what existed. They could only ever imagine the next tiny step. Anything else would have been too much of an abstraction. And that's what must restrict all Idea evolution. If accidents don't happen, the only way an Idea can adapt is with the help of our imagination, and our imagination is, unfortunately, severely restricted.

I'm sitting here trying to reverse-engineer the Ideas behind the wheel and failing dismally for that very reason. The Ideas behind this wheel are too far removed from my experiences; they are out-of-reach abstractions. And in exactly the same way, the other wheel Idea, the one that we all know, was out of reach for the Plains Indians before Europeans rolled up in wagons fitted with them. A Plains Indian upon first sighting a French trapper sitting atop his wheeled wagon may well have thought, just as the eminent Victorian biologist Thomas Henry Huxley thought upon hearing Darwin's theory of natural selection, "How extremely stupid not to have thought of that!" But had he thought in that way, he would have been too hard on himself: the pool of memories in his culture was simply too far removed from the other wheel Idea for him to have pulled off such a feat of genius. There were too many tiny steps that had remained untaken. His culture hadn't even started that particular flip-book.

Life's Ratchet

Ads returns from his solitary walk and we follow the track back to the car and resume our journey to the other side of this cold, green hanging plateau. A few miles on, the enormous basin far below looms into view. I slow down instinctively, because it feels like we're coming to the top of a roller coaster. At the point where we meet the edge of the Precambrian rocks, the Chrysler gradually tips forward, we stiffen in our seats, the dome of the mountain gives us increasing momentum, and we begin our plummet to the present. Trucks coming up the hill bellow but barely move. I, on the other hand, have to repeatedly pump the brakes just to keep us from hurtling down the mountain.

Over the last 3.5 billion years, Life built up momentum just as Ads and I are doing now. The rocks that we fly above hold the evidence. Below us, withstanding our whizzing tire treads, are those ultimately simple single-cell beings trapped in their tombs, but as we drop over the edge of the western escarpment, the rate of advancement in Life's complexity, its rise within design space, is just as nauseating as the rate of our descent. After a long, lagging start, Life ran away with itself. Its exploration of design space got quicker and quicker and quicker over time. Fate, on occasions, "pumped the brakes" by sending asteroids, colliding continents, poisoning the planet with noxious fumes, and sealing off the sun—but to no avail. Life's momentum made it unstoppable.

What drove this feverous progress? What mechanism had the power to lift Life ever up, at gathering speed, through design space, through increasingly complex Life forms? It wasn't a miraculous skyhook—no god was needed. It was a mechanical lifting device, a crane or, more specifically, a "ratchet." Ratchets are gadgets that enable movement in one direction only. Standard car jacks operate on a ratchet. You pull down on a lever. A "pawl" moves over the rounded side of a tooth on a rack of upturned teeth and then falls back onto the topside of the tooth, unable to fall any farther because of the pawl's shape. Hence, with every pull on the lever, the pawl climbs up the rack. There is no danger that it will fall back to ground level, as long as the ratchet is in a stable position on the ground.

Evolution is like this. Natural selection forces the accumulation of adaptive innovations, and each innovation works like a tooth on a ratchet. Once invented, the innovation cannot be uninvented; the pawl is in place and all descendents of the innovator will be blessed with that breakthrough innovation, unable to drop to ground level. So, over Life-time, the gadgets with which new organisms begin their lives become more and more impressive. Just look at us. We come from the factory with the following features as standard: the pyrimidine nucleotide, the single posterior flagellum, multicellularity, bilateral symmetry, entero-coely, the notochord, the spinal column, the jaw, the four-footed body plan, the mammary gland, live bearing, and the placenta. Each of these innovations was cutting age at some point in time, available to only the latest models. Now they are old hat, prerequisites for our design for Life. And it is this accumulation of design that has enabled Life to move ever upward. Each innovation lifts the pawl higher in design space, farther up the ratchet, enabling Life to reach higher still.

Now—and here's my angle—if you cast an eye (itself a pretty amazing tooth) across the ratchet of Life, you'll spot that there is a set of teeth that are rather odd, different from the others. They don't qualify as "in-novations"; more like freak events. Yet, between them, they've propelled Life farther upward and outward in design space than perhaps all the rest. Examining these teeth is in my interest because it will serve to upset the standard notion of the word *organism*.

Building a Super-super-superorganism

SUPERORGANISMS

The first of this odd set of teeth lies at about the two-billion-year mark, or just at the point on our descent when Ads and I are wondering whether the Chrysler's brake lines will hold. At this point in Life, the biosphere is composed almost exclusively of things called prokaryotes, single-cell organisms in which the cell is so primitive that it doesn't even have any internal organization. They are literally just DNA floating about in a water droplet bound by an oily membrane. It's not much of a Life, but it's a living. You'd know the prokaryotes by their catch-all brand name: bacteria. They are a wide and diverse group, ranging from organisms that use photosynthesis to capture energy, such as the cyanobacteria; to organisms that can survive in extreme environments, feeding on nothing but raw chemicals; to organisms that live in cracks in rocks two miles below the Earth's surface; to organisms that now live in your gut and feed on your bran flakes.

Prokaryotes had the monopoly on Life on Earth until a new single-cell organism evolved that had bubbles of oily membranes on the *inside*, too. These organisms could use their internal oil bubbles to store and organize things: special proteins, special fats, chunks of carbohydrates, important minerals, and, crucially, in the largest bubble of all, their DNA. They were the eukaryotes, single-cell organisms *with* internal organization.

One subgroup of eukaryotes was to become especially prominent in the Precambrian world. It was bacterivorous: it ate prokaryotes by surrounding them with blobby arms, engulfing them, and then breaking them up once inside. It had a special type of oil bubble inside it called a mitochondrion. Every cell had a few dozen mitochondria, and their purpose was to break up the remains of the bacteria that the cell ate to get hold of the chemical energy within. In this task, mitochondria excelled. They were far and away the best gadgets Life had ever invented for making available the chemical energy in food. Eukaryotes with mitochondria could extract nineteen times more energy from their food than those without.

All that mitochondria needed to work was a ready supply of the gas oxygen. Originally, this gas was extremely rare on Earth, but an earlier set of teeth on Life's ratchet had changed all that. The photosynthetic work of cyanobacteria gave off oxygen as a waste product, so by about 2.4 billion years ago, as the Earth started to turn green for the first time, the atmosphere became flooded with oxygen, and it was in the wake of this event that mitochondria were invented.

The obvious question is: How did a eukaryote cell invent such an amazing, Life-changing mini-machine? Well, the thing is, it didn't. The mitochondrion began life as a free-living prokaryote; it was a type of bacterium, a distinct organism charged with sourcing its own food and oxygen and extracting chemical energy all for itself. We know this because mitochondria have their own DNA. Their oily membrane is typically prokaryote in structure, they have bits and pieces inside them that are only found inside prokaryotes, and they even divide and reproduce independently inside the eukaryote cells, like caged bacteria.

So how did a prokaryote creature end up inside a eukaryote creature? Best guesses are that it was either eaten by the eukaryote but never digested, or that the prokaryote was a parasite that burrowed into the larger eukaryote with the aim of living within it to steal a little food and oxygen and get a free ride. However the prokaryote first entered the eukaryote, mitochondria have remained there ever since. It's not a bad life. They're provided with all the food and oxygen they would ever need. They're protected from the nasties outside by their eukaryote sugar daddies. As long as they release enough energy for everyone—and they certainly do that—the system works and, more than that, supports every single subsequent heave up the ratchet.

Because mitochondria are creatures within creatures, biologists refer to them as endosymbionts, literally, internal co-habitees. But I want to flip the focus around and look at the eukaryotes that host them. What do we call them, organisms made up of one organism with a colony of other organisms inside them? They're more than organisms. I want to call them superorganisms.

With a branch of Life now super, sporting its own pet energy catchers, the ascent into design space quickened. At about the point that I first try to apply the brakes on the sheer western slope of the Bighorns, at 1.6 billion years ago, a subgroup of this super Life got even more super by repeating the endosymbiont trick. It ingested but did not digest a cyanobacterium, one of those remarkable photosynthesizing organisms that had previously flooded the world with oxygen. Cyanobacteria perform their photosynthesis magic trick by orchestrating what is essentially the opposite chemical reaction to that of the mitochondria: they capture the energy of the sun and produce food and the gas oxygen. So now a part of Life was eukaryotic with *two* colonies of once free-living organisms inside it. And these two endosymbionts—the mitochondria and the "chloroplasts," as they're now called—danced like Fred and Ginger. One captured sunlight energy and wrapped it in food. The other broke up food to get at the energy when needed. One needed oxygen to work. The other produced oxygen as a waste gas. It was a marriage made in heaven, and it meant that these new, really super organisms (with two of the odd teeth on the ratchet) could carry on their autonomous lives wherever the sun shone and water and oxygen were available.

SUPER-SUPERORGANISMS

The next tooth in this odd set crops up at the point of the precipice. It's known as multicellularity, the move from single-cell organisms to many-cell organisms. Multicellularity required single-cell organisms to cooperate as never before, to work together to divide up the tasks of Life between them. Crucially it meant that of all the cooperating cells, only some would be responsible for passing on genes. So, of all the cooperating cells—formerly independent creatures, remember—most had to sacrifice their influence on the next generation, which for any creature is a big ask.

How did it happen? There are many different suggestions, and since multicellularity probably evolved more than once, any number could be correct. Maybe, say some, an error occurred just as a cell divided, with the result that the "daughter cells" were conjoined. If this error was repeated again and again, a multicellular tumor creature would emerge with identical DNA and therefore a common interest. What if, others say, an error created many nuclei within one cell? Indeed, cells in mushrooms and molds look just like this. Perhaps the cells simply combined to battle a problem together, as we see when groups of individual amoebae come together in times of difficulty to create a mobile, sluglike colony called a slime mold.

However it occurred, multicellularity was a tooth of extraordinary significance: it created organisms within organisms within organisms, super-superorganisms that opened up vast new regions of design space and have dominated the majority of Life's eager explorations ever since.

SUPER-SUPER-SUPERORGANISMS

But there's more. Some creatures have extended the habit of cooperation between formerly independent organisms even further to create yet another ratchet tooth. These are super-super-superorganisms—that is, living things that inside them have living things (multicellular beings) that inside them have living things (eukaryotic cells) that inside *them* have living things (endosymbionts). Examples of these include corals, which are a tight association between an alga and a jellyfish; lichens, which are free-living cooperatives between an alga and a fungus; and the eusocialites I first mentioned when introducing the oddities of the naked mole rat.

Eusocial animals are small animals that live in large colonies and divide up the tasks of defense, foraging, and, crucially, reproduction in such a way that, together, they work like one big animal. In place of organs and tissues, this "big animal" has teams of eager, busy individuals. The small animals come in different shapes and sizes to suit their roles, just like the cells that make up big animals. And, again, just like the cells in big animals, the majority of them are entirely sterile because they don't have "reproduce" in their job description. So it's like the first multicellular animals all over again: some individuals make a sacrifice so that others may make it to the Promised Land. However, in this case it isn't single-cell individuals, but multicell individuals, each with its own tissues and organs, that come together as one. Ants, bees, wasps, termites, and naked mole rats all adopt this approach to life, and the strategy works. Naked mole rats may be rare and obscure, but social insects are everywhere, and fundamental to the Life now being led on Earth. Ants alone may account for as much as a quarter of all the weight of animals on Earth. So eusociality was an important tooth. And the same can be said for all the others in this odd set. Every time organisms have merged accidentally to cooperate and, together, get Life lived, Life has become more super, and huge new regions of design space have opened up. The creatures benefiting from these teeth are not just at the periphery of the biosphere, but everywhere and right at its heart. Every living thing larger than a bacterium is a superorganism, a eukaryote with its incorporated mitochondria. *We* are eukaryotes.* And every living thing visible with the naked eye is a super-superorganism, a multicellular eukaryote. *We* are multicellular eukaryotes, eccentric coalitions of cells that would elsewhere be considered independent organisms. These days, super^n-organisms are standard. The living world we recognize is made of living things that have other living things inside them.

So there are very few genuine "organisms" out there. The majority of the inhabitants of the biosphere may look like individuals but are in fact "organi"—compositions, coalitions, and conglomerates of living things

*That's why our chests never cease to rise; we are locked into an oxygen service agreement signed over a billion years ago between our ancestors and a class of wacky prokaryotes with an unequaled talent for getting hold of the energy in food.

that endeavor to take on life in unison. What's important is that natural selection doesn't care. For natural selection, the intricacies, complexities, and obscurities of the organismal offerings that are represented on Earth are entirely irrelevant. However they are organized, natural selection will find a way of testing their organization. If they pass the test, their essence will be handed on to the next generation. If they don't, it won't.

So I can call a tepee an organism without breaking any rules. Perhaps a tepee is a superorganism, with constituent parts that are found living free elsewhere. I'm sure I've seen that tripod Idea living within other cultural organisms. Maybe the tepee is a "cultural lichen," a loosely fastened combination of species that under the right circumstances makes something more than the sum of its parts. The thing is, my analogy between Ideas and biological organisms isn't under much threat when you come to realize what a hodgepodge biological organisms are. And cultural organisms must be the same. The noosphere must be full of $super^n$-organisms. There may be cultural organisms with four, five, or even six *supers*. Natural selection wouldn't care. It will always find a way to test the composite fitness of $super^n$-organisms.

Flippin' Gulls

With the word *organism* unpacked and repacked for my needs, I'll turn to the word *species*. Remember Mayr and Dobzhansky's biological species concept—that a species is "a group of organisms able to interbreed with each other to produce fertile offspring." What biologists don't want you to know is that it's not that simple. Let me give you an example, from my childhood.

I grew up in Torquay, which is a seaside town in Devon in the west of England. It's a lovely place to spend your youth. The English sun shines more often than it does elsewehere. There are boats to sail and rivers to jump in, and I could go rock-pooling whenever I wanted. There was only one drawback: the herring gulls. Every year, herring gulls would nest on the roof above my bedroom. They produce a lot of mess, it's true, and they raid the garbage, but the major problem with herring gulls is the outrageous calls they make throughout the day, starting early. They begin with a few low-pitched judders to warm up their voice boxes, bleating like squeaky goats. (At that point, I'd moan in my half-sleep, knowing what was coming.) Then, when the time is right (and only they know when that is), they launch into a cataclysmic repeating banshee shriek that, when your ears are only feet below their lungs, penetrates your skull like a wobbling drill. (I love them for it, really; it's astonishing that an animal the size of a pet rabbit can make a call that travels from one side of a bay to the other.)

Also listening to their ranting, at the other end of the roof, were "lesser black-backed gulls"—a much more subdued bird, with a cackle rather than a shriek, who would, later in the day, fight with the herring gulls over my chips. They are easily told apart. Herring gulls wear a shawl of gray feathers on their backs and have pink legs. Lesser black-backed gulls have, well, black backs and yellow legs. The two birds don't like each other very much, so while they did sometimes share my roof, they did so begrudgingly.

Linnaeus himself named the lesser black-backed gull *Larus fuscus* in his catalogue of 1758. The Danish bishop Erik Pontoppidan, a big fan of Linnaeus's auditing of God's creatures, gave the herring gull its name, *Larus argentatus*, just a few years later. Pontoppidan could tell that it was a very similar bird—hence he put it in the same genus, *Larus*. However, he could also tell by the way they fought over his chips, and that they would never breed, that they must be separate species. What Pontoppidan couldn't have known, and what I couldn't have known, and what it appears the birds themselves still don't know is that, despite their different looks, calls, and obvious animosity, they are, technically, *the same bird*.

The reason that none of us knew is because you have to adopt a very wide perspective to see the truth. The genus *Larus*, the gull, encircles the entire North Pole as a string of species that inhabit the long north cliff face of all Europe, Asia, and North America. As a brand of bird, their unique selling proposition is that they nest on vertical rock face, fish out at sea, bully smaller birds to give up their catch, and make a lot of noise. The herring gulls that woke me up every morning are just one of this circle. Across the Atlantic, waking up American and Canadian children, is the American herring gull, *Larus smithsonianus*. These birds look and sound similar to my herring gulls, but they are not identical. The American herring gulls have slightly darker backs, brown flecks on their heads in the winter, and different markings on their primary feathers. Nevertheless, when my birds and those birds meet on the lonely cliffs of Iceland, Greenland, or Newfoundland, they will happily breed and produce fully functional offspring. So those two birds, despite their different titles and different Latin names, are the same species, according to Mayr and Dobzhansky.

Then, at the extreme western edge of the American herring gull's range, there's a Russian gull, the Vega gull (often called *Larus vegae*), which has the job of waking up those lazy Siberian children. It's different again. Its back is darker still and its call is less ridiculous, but these differences also don't appear to prevent the occasional liaison between those two neighboring species either. And, just as before, the American herring gull and Vega gull can interbreed to produce noisy little gulls, so they can't be *that* different after all.

And on it goes: to the west of the Vega gull, waking up children in western Russia, is Heuglin's gull (often called *Larus heuglini*). Heuglin's gull has a dark back. It has a cackle call. Its feet are occasionally pink, but mostly yellow. It mates with the Vega gull to the east and, when it can, with the gull to *its* west. And what is that gull called? The lesser black-backed gull. That's where we complete this immense sweep of the world: in Torquay, on my childhood roof, where two gulls sit next to each other begrudgingly. They can't stand each other. They fight with each other to get children's chips. They differently laugh at each other's different-colored feet and backs. They never interbreed, because they are too different—and are, therefore, by definition, different species. Yet because the gulls to their east and west are not so fussy, their genes are, in theory, able to sweep right around the pole, as we have just done, from one gull to another, to become lodged in the other bird on my roof. There, on my roof, they sat, oblivious to the fact that they were on the extreme ends of a single, stretched, circumpolar gene pool.

Remind you of something? Remember the "dialect continuum" to which the Plains Cree language belonged? Every neighboring dialect could talk to each other, but as a result of accumulated change, at the extreme ends of the distribution the languages become so different that they are mutually unintelligible. It's the same story with the gulls. The genus *Larus* is a biological version of a dialect continuum. What we're seeing here is evolutionary change represented in space rather than time. And when it is represented like that, it's technically impossible to demarcate species boundaries, because there aren't any barriers to the flow of genes.

We can make a flip-book out of this. A new kind of flip-book, maybe:

a horizontal, fraternal, flip-book-*in-space*, rather than a vertical, ancestral flip-book-*in-time*. Now, on the first page of the flip-book we have the herring gull that sat on one end of my roof. On the last page there is the lesser black-backed gull that sat on the other end of my roof. In between, over hundreds and hundreds of pages, are all the gulls that could be recruited to pass genes between these two. In order, each one would live to the west of the one before it, so that as you flipped, you'd be journeying not in time as before, but in space, right around the globe.

Flip away. Watch the backs change from light gray to dark. Watch the feet change from pink to yellow. Watch the primary feathers flash. A flip-book can't play bird calls, but were it able to, you'd hear it change in a continuous, contorted song from a banshee's cry to a witch's cackle. The five "species" I mentioned are represented in there, somewhere, but flipping through, you'd be hard pressed to isolate them. The change in the birds' characteristics is gradual and continuous.

In fact, it looks just like an ancestry flip-book. Had you not known that all the birds were contemporaries, you'd have made the assumption that the book portrayed a run-of-the-mill evolution, because gradual change, be it over time or space, leads to essentially the same thing. And we can take something home from this. If we find ourselves unable to demarcate the species boundaries between gulls in space, won't we also be unable to demarcate the species boundaries between gulls in time, or indeed, the species boundaries within any line of ancestors in any ordinary, vertical ancestry flip-book? Take your own ancestry flip-book, for example. Flipping back through the billions of pages, you would be entirely unable to spot the points at which one species became another. After all, gender issues aside, every one of your ancestors would have been able to interbreed with the creatures on the page in front of it, or behind it. It's not a nice thought, but technically, it's true.

So, when looking back in time, you'll see that the starts and ends of species boundaries are just like the starts and ends of Wyoming's coal trains: impossible to distinguish. And that's because they're not there. There was never a point at which one generation was unable to breed with the previous. Mayr and Dobzhansky's definition doesn't work when looking at changes in living things over time. The best thing you can do is look back after the event and say, "Hey, it's happened!" And, as the gulls show us, it doesn't always work over space either. The two gulls that were on my childhood roof don't interbreed. So they are, according to the biological species concept, two different species, and yet they're not. And the reverse can also be true, as Ads and I are about to find out in Yellowstone.

The Yellowstone Blues

"It's a free day, sir," says the man in the hat at the East Entrance of Yellowstone National Park. He clearly thinks that I don't understand this concept, because he's compelled to repeat himself: "It's a free day, sir. You can drive on in." He's right. I don't, and my hand continues to hold out its twenty-five dollars.

"A free day?" I say, "but . . . why?"

"We do this every now and then."

For a moment, I'm paralyzed by the good fortune. But then I hear Ads sigh; I pull myself together and press the accelerator.

Ads is almost jumping up and down in the passenger seat with excitement. We wind up through a spectacular ravine clad in tall pines and decorated with a cornice of gray cliffs. We summit the pass and are presented with, at this late hour, the best landscape painting you could imagine: Yellowstone Lake, smokey blue, set against a sliver of gold sky and topped by a powder keg of cumulonimbus that shakes out its water over the lake islands in a fine mist. We drop down and sweep around the lake. Stinkpot sulphur leaks send us diving for the buttons to close the car windows. The beaches are empty in the rain but for a solitary bull buffalo. We find the campsite hiding in the cold forest, heed the notice that a grizzly walked through here this afternoon, and pitch the tent. I hand Ads a beer. He finally settles.

I can understand his excitement. Yellowstone deserves its reputation as North America's most exciting natural wonderland. Not only does the terrifying super-volcano that lurks under the earth here operate an award-wining geothermal theme park (resplendent with dozens of geysers, hot springs, steaming vents, and mudpots—no two alike), not only does the Yellowstone River excel itself in its choice of path, away from this elevated bowl (cascading, then tumbling, then roaring, laughing raucously as it carves titanic canyons), but the forbidding height of this plateau on the Rockies and the positive management of the world's oldest National Park have nurtured the closest thing North America has to a complete endemic ecosystem.

The next morning, stuck in a seasonal "critter jam," we sneak, inch by inch, toward the best view of this ecosystem: Hayden Valley. After an hour, the procession of RVs, SUVs, and we's finally gives up on the idea of making progress and turns into a unidimensional car park. I turn off the motor. "This is why it was a 'free day,'" I announce to Ads, in an effort to rescue some glory from the situation. He ignores me and stares out at the little scenery we can see. We are among lodgepole pines, the trees that the Plains Indians preferred to use for their tepee poles. They are strong and straight, with few lower branches, and they're light. With only travoises at their disposal, there was an obvious benefit to a low-weight timber: the lighter the poles, the more you could drag, and the larger your lodge could be. But it strikes me that we are a long way from the plains, and I suddenly get an image of a pilgrimage in which the Cheyenne and Arapaho, Crow and Comanche separately make their way up into the mountains to find the trees that will keep their tepees straight that year. In the absence of the other wheel Idea, their lodgepoles had short lives and needed to be replaced often.

"And this is why we're stuck," says Ads suddenly. I glance back, and right there at the end of our car I see a bull buffalo, a mist of sweat blowing from his clammy nostrils. He's clearly as cross as we are with the rank of cars; in the urgency to go nowhere, we've all ended up nose to tail, leaving gaps between the vehicles that are too small for buffalo to navigate. The way he's looking at me, it's apparent that he thinks it's my fault. He vents his frustration with a few cough-snorts, then bulldozes between our hood and the trunk of the car in front. We hear the metal cry out, and the family ahead are bounced up and down as the huge beast trips over their trailer hitch. It's like a scene from *Jurassic Park*.

With the buffalo gone, the traffic moves again, the woodlands end, and we are met with a wide open grassland that slopes up, away from us. Sprinkled like confetti across the enormous valley is a huge herd of buffalo. They are all sizes and ages. Bulls croak at one another across the plain, huff and puff under their weight in the sun, roll on the ground, blowing up thick clouds of dust. Cows cluster around the creek, arguing over who owns what. Youngsters skip about like kids in a playground. In the distance we spy a small gathering of elk. Beyond, in the forest, are grizzlies and black bear. Moose dawdle in the wet meadows. On the hilltops to the north are packs of wolves that prey on mule deer and pronghorn. There are bighorn sheep on the roof of Yellowstone, four types of rabbit, nine members of the squirrel family, seven types of shrew, nine of bat, nine of weasel, ten of mice, a beaver, a pocket gopher, a porcupine, a raccoon, and a pika. That's just the mammals. If you audit the plants, the invertebrates, the birds, the amphibians, and the reptiles, the biodiversity of the "greater Yellowstone ecosystem" is incomparable in the northern temperate zone.

But its fish are a mess.

While the wild, green lands of Yellowstone have gone from strength to strength, the blue places of the park have witnessed a biological catastrophe. When the park was opened in 1872 the undisputed super-predator of the waterways was the Yellowstone cutthroat trout, a handsome, sparkly fish with dark spots, gold skin, and tasty pink flesh. Initially the impulsions of fly fishermen favored this highwater champ; they helped it up and over the immense Yellowstone waterfalls and into the backwoods lakes, to places it had no hope of conquering by its own means. But subsequently, this native has suffered the invasions of four exotic trout species at the hand of the same, insatiable fishermen: brook and brown trout from the East, lake trout from the North, and rainbow trout from the West. The Yellowstone cutthroat can't compete with any of them. The brook trout are hardier in the wintertime. The brown trout is bigger and better at harvesting the fly hatchings. The lake trout preys directly on the young cutthroat. But the rainbow trout is perhaps the most insidious in its attack: while the other nonnatives steal the Yellowstone cutthroat's habitat, prey, and young, the rainbow trout steals something much more precious: its genes.

It turns out that, although rainbow trout are a different species from the Yellowstone cutthroat trout—from a completely different part of the world—no one has told *them* this. The two fish species merrily interbreed whenever the chance arises. The resulting hybrids, known as cutbows, are far from impotent "mules"; they are highly fertile, viable fish that show a degree of dominance over both their parent species. For fans of genetic purity, this is a disaster. The endemic Yellowstone cutthroat is disappearing before their eyes, their uniqueness swallowed in fresh generations of mongrel fish. But it's not that great for fish taxonomists either. In vigorously defying Mayr and Dobzhansky's biological species concept, the trout are making a mockery of the taxonomists' efforts to catalogue them. Hybridization steals the one objective measure upon which the whole of their work rests. How do you classify a cutbow? Do you list it under "rainbow" or "cutthroat"? One solution might be to pull back and conclude that, despite their dispersed origins, the rainbow and the cutthroat are somehow one species. But even if we did this, the "species problem," as it's called, wouldn't then go away. Although biologists hate to admit it, hybridization—the polar opposite of speciation, where *two* species become *one*—is rampant throughout the lower reaches of the tree of Life. The big animals that Ads and I snap photos of in Hayden Valley may play by the rules, but smaller vertebrates, plants, invertebrates, and, heaven knows, the mass of simple organisms that swarm at the bottom of the evolutionary tree, routinely spill and mix their gene pools. Down there, it's a genetic quagmire, a soggy mess in which the word *species* has very little meaning.

Life Is Simple

And it gets worse. In addition to this forbidden interbreeding, we're finding that living things further subvert their "species" name badges by casually trading small sets of genes willy-nilly, with completely different species, in a practice known as horizontal gene transfer (HGT). Bacteria especially will habitually inherit strands of foreign genes from other types of bacteria, either via mischievous viruses, or upon absorbing vagrant genes through their cell walls by accident, or by actively puncturing each other's cell walls and swapping their DNA. This explains a phenomenon, first witnessed in 1959, whereby bacteria of many different species simultaneously develop resistance to an antibiotic. They're not individually evolving their resistance; they are passing the gene that does the job across the species barrier, many times.

So the tree of Life is not the handsome standard we once believed it to be. It's a knotted coppice, clumped at its base like the mycelium of a fungus. And the closer you are to the base of the trunk, the more hellish it is for the taxonomists. In Yellowstone National Park, you can visit this taxonomists' hell by crossing a little trip-trap bridge across the Firehole River (and its resident cutbow trout). The other side looks as if it's already been stripped by the billy goats gruff. There's no grass, not even soil, just a dome of bare rock caked in silica spills. As Ads and I step onto this dome, we effectively cross into the Archean: the Earth two billion years ago. Apart from the long line of tourists, there are no multicellular eukaryotes here. It's a unicellular world with an angry, primordial god at its center puffing out a fury of steam: the Grand Prismatic Spring, the largest hot spring in North America. Boardwalks lift us safely toward its mouth, over the crusty, mucky hill it has spent the millennia building. The rock below starts to glisten with a psychedelic prokaryote slime. It turns red then orange then yellow then green. Each band in this microbial rainbow is a mini-ecosystem, a bacterial mat forest with its own mini-canopy and mini-understory of cyanobacteria lying half an inch above its own mini-forest floor. Each time the cyanobacteria change, the color changes. The closer they are to the hot mouth of the spring, where the water is lifeless and ice blue, the better they are at taking the heat. The yellows sit happily all day in temperatures that would melt your skin.

This is the sort of ecosystem that once covered all the wet surfaces of the planet. And it's the sort of place in which the illegal activities of hybridization and HGT are pursued in abandon, creating the sort of Life that defies our categorization. Yet in these dissident bacterial forests, Life is still lived. Evolution continues. Natural selection again does not appear to care if Life chooses to abide by our idealized definitions. After all, its job is not to judge the conformity or otherwise of living things to our preferred title "species." Its job is just to test their fitness.

The same must be true in the noosphere. Any of Life's offerings there will be subject to natural selection, just as here. So I can call a tepee a species if I want to, because the word *species* doesn't mean much. If I ever do find a tepee with both a tripod *and* a smoke flap hole, so what? It's then a hybrid, a cross between two other tepee species. There are plenty of fully operational hybrids at large in the biosphere. Why should it be different in the noosphere?

I don't need to get hung up on the words *organism* and *species*. They are terms that we invent to categorize Life for our own convenience. The names of gulls, of bacteria, of trout, and of lichens are fabrications. The gene pools of these species are not bound by anything real; they're bound only by our imaginations.

A rainbow flashes hesitantly in the steam above the springs. Ads and I can both see its seven bands of color. But in fact, they don't really exist. Rainbows are optical phenomena in which the entire visual spectrum is offered to the eye. We're looking at a continuous spectrum of light. It is only as a consequence of the way in which our visual system interprets the incoming light that we "see" the stripes of color. And we must recognize that we have the same problem with the biosphere. The living world is a continuous spectrum. The boundaries we "see" are more often than not acts of our imagination. Genuine divisions and border lines are rare in the living world, and *natural selection doesn't need them anyway.* Evolution still happens no matter how many rule breakers there are.

It was in facing this truth—the reality that the words *organism* and *species* mean less and less the more you study Life's explorations in design space—that some biologists were brought to question their whole worldview in the 1960s. If, they asked themselves, the two most important words in our vocabulary are essentially worthless, then where are we left, and, more important, what are we left with? William Hamilton and George Williams, two prominent evolutionary theorists, independently came up with the answer. And the answer, rather unappetizingly, was this: nothing but genes.

Just as the only real substance behind a rainbow is a collection of photons, the only real substance behind biological Life is a collection of genes. If you cast a critical eye at Mr. P. Veale's levels of organization rank, sore from the realization that the word *organism* is a troublesome term, then everything ultimately reduces down to the level of organization on the extreme left: the genes.

Hamilton and Williams had given their science a new pair of goggles: gene's-eye-view goggles. Look through these, they would say, and suddenly you see Life uncluttered by the fabricated order we attempt to place on it. Look through these, and you see Life is simple. It has only one purpose: to do anything and everything to help the genes pass to the next generation. Whatever best achieves this aim—no matter how complicated, wacky, involved, or "illegal"—will be promoted by natural selection and adopted by Life as good design. And now, they would tease, take a look at evolution. Through these goggles, it's nought but a melee of competing genes desperate to sustain themselves, by whatever means. The ratchet at its heart is an accident of their approach. The ascent of Life up through design space is merely an epiphenomenon of their eternal efforts to achieve immortality. And the driving force of Life, Hamilton and Williams would say, finishing with a wink, is the unrelenting, inexorable selfishness of the genes.

"My turn to drive," says Ads, snatching the Chrysler key from my hand. "Fine," I huff faux-fully, "it's all about *you,* Adam, after all." Now

Ads will be the one at the wheel as we follow the Yellowstone River out of the park and north back into Montana. He'll have the joy of weaving through its canyon and on into the wide open valley beyond. Selfish bastard.

He does just that. He leads us down from the high plateau, past more bubbling silicate terraces at Mammoth. He beeps at the elk that stand on the road eating a hotel's shrubbery. He guides us out of Wyoming, out of the park, and into Montana.

If it weren't for the signs erected at these points, we would have no way of knowing where these boundaries lie. They are superphysical, forced, arbitrary borders, notionally drawn (by men in top hats, I assume) as straight lines across the unruly landscape. They look real enough, from a distance, on a map, but when you're there, up close, they don't exist. Like the organisms and species of the prokaryote world, Wyoming, Montana, and Yellowstone National Park are bound only by our imaginations.

As we cross (apparently) into Montana, the Yellowstone River skips ahead, enjoying the space of the wide valley it has carved. It sidles at speed toward and away from us, stretching out, playing with the mountains that rise on either side, snipping off a bit here, a bit there, to carry as a souvenir all the way to the Gulf of Mexico. And it is then that I realize that my new world view needs its "gene." If cultural Life is like biological Life, it, too, must have an underlying substance, a fundamental unit. These goggles, in many ways the daughter of the goggles that Hamilton and Williams created, must be powerful enough to isolate that stuff, to root out the "genes" of culture, to identify the selfish bastards that must surely, even now, be enjoying themselves, seated at the wheel of cultural Life.

13

The Genes of Culture

A Model Idea

To have some chance of identifying the "genes" of culture, we need to find ourselves what laboratory biologists would call a model organism, a simple, familiar, convivial living thing to scrutinize. Famous model organisms of the biosphere include *Escherichia coli*, the human gut bacterium; *Caenorhabditis elegans*, the minute roundworm; and *Drosophila melanogaster*, the fruit fly. In our case, we don't want a bacterium or an animal species, but a species of Idea. So what could be considered a simple, familiar, convivial Idea?

The essential word here is *simple*. We are so proficient at handling Ideas that we find many Ideas simple, but that doesn't mean to say that they are simple species of Idea. The Idea of a "coffee shop" is a simple Idea, one would think. Yet it requires all sorts of other Ideas to be in place before it can survive successfully in the mind. You need to understand language, for a start; then be familiar with the words *coffee* and *shop*; then understand the point behind putting them together, and so on. That's not simple. That's not an example of a cultural prokaryote, equivalent to *Escherichia coli*.

I've got an idea. Humans aren't the only Dennettian creatures; there are other animals out there that dabble in a spot of culture. Like us, they all tend to hang around in groups, have time on their hands, possess good- to middling-size brains, and show an inclination to imitate. Presumably, because they are much thicker than we are, the Ideas that they work with must be about as simple as Idea Life gets. So what sorts of Ideas do they host?

Deep in the Congo, a smattering of chimpanzee troops have discovered the Idea of using clumps of leaves to hold water like a sponge as they walk through the forest. We know this is a genuine cultural behavior because the chimps don't use clumps of leaves like this instinctively; each new generation must learn the trick from scratch by watching their elders. So the behavior is not passed down in the genes: it's not hardwired: the information transfers culturally, and lives on only in the chimps' memories. If, on one unfortunate night, the tree that the whole troop was sleeping in happened to tumble, and all the chimps hit their heads and lost their memories, the clump-of-leaves-like-a-sponge Idea, and the rest of their culture, would be lost forever.

Elsewhere, there's a troop of chimps that don't know about the leaf trick but that have a different Idea: they've worked out how to extract termites from a nest with a fashioned stick. Still other troops of chimps have no Idea about those two gadgets, but they happily teach their young how to open a hard nut with a "hammer" stone on a larger "anvil" rock. So chimps do have cultures, and aside from us, they are probably the most accomplished Dennettians on Earth.

But it's not just chimps. Japanese macaque monkeys have learned to throw rice into seawater to separate the water from soil and sand. Orangutans have watched unsuspecting humans and learned how to row boats. Orcas in the shallows of Patagonia teach their young how to beach themselves to catch fur seal pups. Dolphins off the west coast of Australia have learned to balance sponges on their noses for protection when on a stingray hunt. Then there are the elephants that teach one another to avoid land mines. None of these behaviors are performed instinctively; they all must be learned and retained in the memory for future generations.

So the common theme here is that all the other Dennettians appear to be restricted to indulging in what we might term "psychomotor Ideas," Ideas based upon the use of body muscles to copy physical behaviors and nothing else. No words are needed to pass these from mind to mind; you just watch and learn.

If psychomotor Ideas are the simplest of all Ideas—the prokaryotes of the noosphere—then it's not hard to find a familiar, convivial species to study. Take the act of tying a shoelace. Virtually all of us host this Idea in our brains, but there must be thousands of shoelace-tie species out there, there must be thousands of ways in which a shoelace can be knotted together to keep a shoe firmly on a foot.

I tie mine in a bow (with a double knot, unless I'm likely to need to keep slipping my shoes on and off). I make a simple over-and-under pass, then collect one end of the lace into a loop (which takes both hands), wrap the other end around it (the so-called "round the tree" action), and then pull both loop heads into a bow, evening out the laces until they look tidy. I do the last bit again if I'm keeping my shoes on for a while. It's not easy to explain in words *because* it's a psychomotor Idea—which is why the best way to inherit the Idea is for someone to show you how to do it.

My mum showed me. She had to be patient, I was about four. She probably taught me to do it because I was going to school for the first time, I can't remember. But I can remember the effort that it took to learn the technique. I suspect that I had my tongue poking out like four-year-olds do when they're really concentrating on an under-rehearsed motor task. I remember trying again and again, my hand slipping and losing the knot repeatedly, simply because I couldn't control the action well enough. I hadn't had enough practice. My clumsy finger buds were a bit rubbish.

But my brain must have been whirring like a LAN file server. Imitating something you've never done before is an astonishingly sophisticated computation because it involves transposing the actions of another individual. The ligaments and muscles in my arms and hands were replicating the actions of the ligaments and muscles in my mum's arms and hands. I was converting all that hidden mechanical movement in her body into mechanical movement in mine—and that's no picnic. If you don't believe me (and you've already conquered the shoelace-tie Idea), try learning the foxtrot.

In addition to replicating those movements, I also had to understand the causes and effects of each of her actions so that I wouldn't copy irrelevant ones. For example, if my mum's patience was waning (which it must have been), she may have flicked the wrist I was staring at so intently to glance at her watch, mid-demonstration. My four-year-old brain, even that baby brain of mine, knew enough about the world to understand that the wrist shake she did, mid-demonstration, was nothing to do with the shoelace-tie Idea. When I next tried to tie a shoelace, I wouldn't have flicked my wrist at that point, because I would have understood that it wasn't a causative component in the act of shoelace tying. This cause-and-effect intelligence must be a crucial tool in our Dennettian boxes.

So learning how to tie a shoelace is a huge achievement, and it results in something quite remarkable: by the end of the lesson, with my new skill sussed, my four-year-old brain would have built a facsimile of the instructions that my mum had in her head. My intense concentration and practice, and her persistent tuition, had effectively enabled me to copy a piece of her memory. It's as though I had downloaded wirelessly the hundreds of programming lines in the shoelace-tie-with-a-bow software package that her mind was hosting. I had inherited an entire intact psychomotor routine (Idea) fragment by fragment.

Those fragments of information that were cut and pasted wirelessly from one mind to another must be the fundamental units of the Idea, the "genes" of a cultural species. I'm starting to realize that that's what these goggles are built to see.

Blackfoot Country

The mess of strip malls and chain stores running through Great Falls in northwest Montana has been the worst yet; "there's no *town* in that town," said Ads, and he was right. The taco dispensers, drive-in casinos, and diners-with-a-live-mermaid-show never seemed to end as we dissected the conurbation, and it took us miles and miles to shake them off our tail, but now we are relieved to find that the plains have returned with a vengeance. If anything, the shortgrass here is more desolate than any we've seen on our trip. We head northwest, out into the heart of it. The sun's getting low, blinding us from the west and casting the whole scene in a golden brown. The place is dry, crisp, recovering from a hard, hot day. Stark, upright hills break the backdrop on occasion, their dark, shadowed faces staring at us as we pass. The latent heat from the grass blows at us through our open windows.

Then, with nothing but an enormous bowl of deep blue above, we hear the sound of heavy raindrops on the windshield. I hang my hand out but can't feel anything. The shower suddenly gets stronger—splat, splat, splat—it sounds like lazy summer rain, but looking ahead, we see yellow spots and streaks start to paste over the glass, and tiny brown limbs and paper wings splayed at all angles. It's raining bugs. Ads puts the wipers on, and they groan as they clear the dry windshield of its insect carcasses and deposit them to one side. The shower gets thicker. We wind up the windows as bugs start to hit the backseat. The noise gathers until it sounds like a hailstorm. Ads turns the wipers up a notch. They flap back and forth impatiently, parting each wave of attack. A mass grave of insects accrues on my side of the windshield. Each swish brings a new load of bodies, and the sun appears to take some sordid pleasure in highlighting the gathering thousands of yellow flecks that mark our many murders. Ads hits the washers.

The insect cloud disappears as we leave a hilltop and drop down toward the next coulee. The wash rinses the insect stains away, enabling us to take in the broadest landscape either of us has ever seen: an enormous sweep of dark, gold grass, inestimable in its span, gathered in bank after bank of low ridges so that it looks like an ocean of yellow rollers is coming our way. The sky above is like that on a child's painting: solid, blue, and taking up much more than half the frame. But squeezed between these two giants is a ghost: the spirit of the Rocky Mountains, hanging over the horizon, suspended somehow beyond the grass, in a long, jagged line of specter gray peaks.

We've reached the Rocky Mountain Front. It's aptly named. Farther south, where the Great Plains and the Rockies meet, the encounters are dramatic yet cordial. Here, in northern Montana, the two geographies look like they're at war and this is a line of engagement. It extends for more than five hundred miles, from just to the south of here to up into northern Alberta, Canada. Along this length, the mountains lose some territory to the plains, but at this position the mountains are proud and dominant and the Great Plains ruffle backward in response.

This is without doubt the toughest part of the plains: a felted moorland, dry and baked in summer, frozen in winter, whipped by the Chinook winds, exposed, at high altitude, and up against the mountains. It's utterly desolate. The shortgrass looks like it takes a beating every day of its life. It's stumpier, oilier, and yellower than ever, its heads bowed over as if its spirit has been broken.

The Indians who called this part of the plains home when the Europeans first visited were given the label Blackfoot. In fact, they were not one tribe but a tight association of three tribes with a common language: the Piegan, the Blood, and the Siksika, the latter name meaning "those with black moccasins" in their neighbors' tongue. Surrounded on all sides by enemies, the Blackfoot Confederacy were fiercely protective of their fierce territory. The Blackfoot name, more than any other, would strike terror in the minds of European explorers and trappers. And in the end, it was the loss of their buffalo and the coming of smallpox, rather than the armies of the United States or Canada, that would break their resistance.

In defining the United States and Canada, government officials cut Blackfoot country in two by one of those superphysical, cultural borders. Today the Blackfoot live as the South Piegan in northwest Montana, and as the North Piegan, Blood, and Blackfoot in Canada. The bulk of their land is in Canada. Their American enclave is run from Browning, Montana. The town comes into view as the sun sinks below the ridge of the ever-present Rocky Mountain Front. Its lights flicker on as we approach. Browning looks like it's just been tipped out of a box onto the plains. Flimsy wooden buildings litter the wasteland at all angles. There's no hurry for a town center.

We pull into a gas station that turns out to be the local supermarket, teenage hangout, and stray dog haunt all in one. I step out of the car and nearly trip over two dogs humping, the bitch spinning her mate around in a vain attempt to slough him. Inside, dozens of youths with long, shiny black hair, dressed in vests, basketball shorts, and running shoes, traipse about the aisles, independently and in silence. I lean over several to gather snacks, then settle up and get out.

We attempt to put some miles between us and Browning, but soon I spy silhouettes of tepees on a hill next to a small house with a veranda. There's a sign on the road at its gate, "Lodgepole Gallery," so I veer off without much warning, explaining to Ads that we may still bag a new tepee before the day is out.

We park the car at the top of the drive. In front of us is a magnificent tepee, proud on the hilltop. The main body of the canvas is painted sun yellow. The top has a band of sky blue, with large yellow spots on the smoke flaps. At the bottom is another blue band, scalloped and spotted with yellow. About halfway up there is a single thin blue stripe, and standing on top of that, a line of identical blackbirds, their heads hung downward as if in prayer. This is the first fully painted tepee we've seen. At the Crow Fair, virtually all the tepees were white. There were a few with red or blue bands at their peaks and the odd icon, but none was covered with paint like this. Here, the painting is not just symbolic, it's joyful, and the joy is so powerful that it feels like a statement.

I do a quick survey of the tepee's structure. If this is genuine Blackfoot, then they were, as Kenny said, a four-pole people. And guess what: the smoke flap poles stick through holes and are secured with a cross-stick, just like the Crow tepees. The smoke flaps themselves look stubby compared with those on the other tepees I've seen: short, so that the distance between their lower edge and the doorway is greater, requiring a larger number of lacing pins to tie the two ends of the cover together.

I walk toward the homestead, shouting out to announce our arrival. The owner calls from inside, welcoming us into his gallery. There are paintings of plains life, crafts, and artifacts throughout the room. The owner, Darrell, is one of the artists, and he quickly returns to a painting on an easel while we circulate. After an acceptable period of browsing, I ask him if Blackfoot tepees are always painted.

"Not always," he replies while dabbing at the canvas in front of him, "but we paint our lodges more than any other, and we have our own way of painting."

I ask him how the Blackfoot tepee is different from the Crow.

He stares at the ceiling, then offers: "Well, we don't go for those ties that come out in front of the lodge."

To me, this seems a trivial difference, so I want to press further, but I'm conscious that he's keen to continue his work. I do another circuit of the gallery, then ask about the birds on the painted tepee outside.

"They're crows," he answers, then adds with obvious satisfaction, "that's one of only a hundred and eighty medicine tepees belonging to the Blackfoot Confederacy."

Of course I have to ask: "What's a medicine tepee?," at which point, Darrell jumps to his feet and darts off into a back room filled with paint pots and brushes. He returns with a book—*The Indian Tipi*, by Reginald and Gladys Laubin, written in 1957—slaps it in my palm, looks me straight in the eyes, and says, softly, "In here are all the answers to your questions," before sitting down at the easel again. An hour later, by LED light in our tent, I discover that Darrell was absolutely right.

The Idea Behind These Goggles

"Who's driving?" I ask.

"I don't mind; you can if you want," says Ads.

"No, why don't you. The drive today is going to be amazing."

"So why don't you, then, if it's going to be amazing?"

"Okay, I'll tell you what, I'll drive one way; you drive back."

"Okay," he agrees, then, after a moment, adds, "So, who's driving?"

It's high time that I explained how my goggles were made. As I mentioned previously, these goggles are the work of an assorted bunch of thinkers from a number of different academic disciplines, who don't always agree, but they're united on one point: if culture is evolving, then *something must be driving it*, because otherwise it wouldn't go anywhere in Life and, in fact, would not even have started the journey it's on.

The goggle makers drew this conclusion only after adopting Hamilton and William's gene's-eye view, so in describing the goggles' construction, I must first describe this:

Remember, back in South Dakota, I said that Life, in order to cheat death, invented genes so that instructions on how to make organisms could be carried on to the next generation? Well, the gene's-eye view flips this explanation on its head and suggests that it was *genes* that invented Life in order that *they* might cheat death.

Why did they need to invent Life? Genes are replicators, entities built in such a way that they are able to make copies of themselves. In theory, a replicator can go on forever, copying itself, so long as it is provided with the right materials and the right environment in which to work. Genes built living things (starting with the simplest prokaryote*) in order to provide for these two needs. But the downside of building a nice living thing in which to live is that natural selection will test that living thing, and if it's poorly built—unable to survive and/or reproduce—it will perish, along with the genes that built it. Hence, naturally, and accidentally/automatically, it came to pass that only the genes able to build the most successful living things stuck around.

But what exactly do we mean by "successful"? Well, that's where the gene's-eye view comes into its own. Natural selection, it says, cannot deal directly with the replicators. The only things that natural selection can "touch" are the *effects* of the replicators, the characteristics that the replicators are responsible for manufacturing in living things. In the language of the gene's-eye view, these characteristics are known as interactors, because it is they who interact with natural selection.

Thus, at the very center of all Life, a cozy deal has been struck between replicator and interactor. Replicators that accidentally/automatically manufacture interactors that assist in their voyage to the next generation will inevitably become more common. Those that don't, won't. End of story. The living things themselves, which are no more than composites of interactors, are cut out of the deal. They don't get a look in. So the word *successful* in the earlier sentence means "from the gene's point of view." If living things must be sacrificed in order to bring the greatest net benefit to the replicators inside them, then that's what will happen. Bees die after stinging, salmon die after breeding, Florida scrub jays remain celibate in order to help their parents raise a new clutch, and naked mole rats remain sterile in order to work for their queen. Life is full of sacrifice—individuals paying the ultimate price in order to benefit the replicators that built them. It looks, for all the world, like the genes are "selfish," but of course they can't be; they're not sentient beings; they're molecules. It's just the way that Life's balances have been set that gives that impression.

But the genes don't have it all their own way. They're tied into this deal as much as the others. And because the world's resources are finite, and increasingly hard to acquire, the genes have found themselves trapped

---

*Actually, we don't know what they started with; even the simplest prokaryote is a hell of a jump from nothing. It was probably something simpler still—a virus perhaps—which subsequently became usurped by the simplest prokaryotes, but the jury's still out at the moment.

inside their various brands of living thing engaged in a maddening arms race. Every living thing that they manufacture must be better designed than the last, able to compete with the others in their species and with the others in *other* species in the wider world. So there is a relentless clamber for betterment—and it is *this* that powers the drive shaft of Life. Biospherical Life looks as it does only because it is replicator-driven.

Realizing this, the makers of my goggles* had an Idea. They concluded that since cultural evolution looked so similar to biological evolution, it must also be driven by its own underlying replicator-interactor relationship. But they knew that the replicator in this relationship couldn't be the gene because, as I said in chapter 1, so many of the things that we do in the name of culture either hold no benefit for the genes or actually hinder them in some way. So if culture wasn't gene-driven, then what *was* driving it? In 1976, the chief publicist of the gene's-eye view, Richard Dawkins, declared that the only conceivable answer was that culture must have its own, new replicator. An entity with its own interests. A unit that copied itself by using not our cells but our minds. A body that secured its passage into the future only by building successful Ideas that could leap from one mind to the next. He called this new replicator the meme.

The fragments of Idea that were passing from my mum to me as she begat her shoelace tie were these genes of culture, these memes. Each of the hundreds of programming lines that I downloaded wirelessly and then tucked away in my memory when learning to tie my shoes was one of them. Once housed in my memory, these memes would manufacture their interactors, the various psychomotor behaviors required to tie a shoelace in that particular way, the many bits that make up the Idea. And, according to "meme theory," it is the struggle between these replicators to secure a position in the collective human repository that drives cultural evolution.

Let me give you an example. In my own household there's a struggle for existence between two rival shoelace-tie Ideas. There's my shoelace-tie Idea—the one with the bows, which, looking around, seems to be the dominant species in Southern England—and then there's the different, closely related species that harbors in my daughter Mia's mind, an Idea that stems from the Netherlands. When Mia was four, her best friend

*See the bibliography, "The Goggle Makers."

was a Dutch girl. One afternoon, for something to do, they were both taught to tie their shoelaces by Mia's friend's mum, who ties a shoelace by making two independent loops first and then lashing them together to make a bow; no "round the tree" for her. The method seems a bit clumsy to me, but my daughter much prefers it, and is now the happy host of an immigrant species of shoelace-tie Idea.

The difference between my English shoelace-tie Idea and my daughter's may be down to only a few memes and their interactors—a slight twist of the hand here, an extra loop there—but that difference creates two discernibly different shoelace-tie Idea species, and they are compelled to compete with each other for space within the finite environment of my family's collective memory. I know Mia's shoelace-tie Idea, so the memes are in my memory, but I only ever use the one I've always used. In me, those Dutch shoelace-tie memes are sterile; aside from mentioning them here, I won't be helping them to spread to any minds. Why do I prefer my Idea over hers? That's a good question, and one I'll have a proper go at before this trip is over, but for now suffice it to say that, ultimately, cultural evolution comes down to this kind of meme tussle. Only those memes that work in unison to build the most successful Ideas will stick around, with the understanding that, again, the word *successful* means "from the replicators' point of view." Cultural evolution is tainted with the same replicator selfishness as biological evolution, and the living things of culture, the Ideas themselves, will be sacrificed accordingly.

And what about us, the hosts of this meme Life? If memes will sacrifice their own living Ideas, then what will they do to/with us? For our species, the most important revelation of the view through these goggles—this meme's-eye view—is that in adopting culture, we have the dubious distinction of becoming subject to the activities of not one, but two "selfish bastards." And that, as we shall see, explains everything.

The Indian Tipi

The Comanche used a four-pole base that actually looked like three-poles . . .

Assiniboine tepee poles spiraled in reverse of Sioux and Cheyenne . . .

All four-pole people use smoke flap holes, except Comanche and Shoshone, who use pockets . . .

Crows use a special tie . . .

Cree have different smoke flap pockets to other three-poles . . .
The Kiowa—every fourth tepee was painted . . .

With the brilliantly lit, battered gray torso of Glacier National Park looming over me, its legs hidden in a skirt of shimmering pine and aspen forest, and a black bear wandering behind me somewhere about the lake, I alternate between jotting notes from *The Indian Tipi*, clicking links on the laptop, biting neat scalloped edges in donuts, and sipping too-hot coffee. I'm in heaven. The clues to the origins of tepees are hidden among these paper and digital pages, and I feel like I'm solving a detective novel a hundred pages before the end.

As I flick through my new book, dozens of illustrations rich in information spin by: drawings of smoke flap shapes, of tribal knots and ties, of hide patterns, of doorways and medicine bags—all busy with labels like the sketchbooks of nineteenth-century naturalists. And in among those illustrations flash even more precious, old, fuzzy black-and-whites, relics of the late 1940s and early 1950s. They show hands and boots on the edges of frames and enigmatic cowboys, their faces hidden in the shadows that live beneath the brims of their hats, demonstrating how to erect a three-pole, how to make parfleche (rawhide), how to prepare a ground oven. And then I find, secreted in the very center of the book, a trove of color photographs. Waxy, low-contrast, and dull, they show the insides of a tepee decorated with an attention to detail that is too exacting, too enthusiastic, unreal: the labor of two people desperately in love with another people's culture. In the middle of the middle photo, in the middle of the book, are the two people in love: Reginald and Gladys Laubin. They sit in the middle of their authentic Sioux tepee, Reginald leaning against his authentic Sioux backrest, Gladys leaning against her authentic Reginald. Both are dressed in authentic Sioux natural-dyed clothes, wearing jewelry, headdresses, and face paint—pretending to be Plains Indians, gazing inauthentically to the left of the lens, their eyelids half-closed, pretending to be peaceful, like the subjects of a Raphael.

Sixty-odd years ago, these two were busy people, a white American couple frantically trying to reassemble the "old way" of the Plains Indian before it was forgotten forever. Even in the fifties, less than a century after the tribes left the plains for their new reservations, the Ideas of Plains Indian culture were in critical peril; they lived on, lost and unused, at the backs of only a very few minds, jumbled up with unmarked mutations. Repeatedly in the book, the Laubins explain that the information they gathered from members of different tribes on how to make and erect the different tepees could not always be trusted. With their mistakes and inconsistencies, the young and often the old Indians would demonstrate just how much the tribes had forgotten about their tepee days. The Laubins did the work of sorting the wheat from the chaff, and then, after years of hunting and gathering the authentic memes, they set to ink, forever, those disappearing Ideas. If it hadn't been for the Laubins, the collective repository of America might have lost much more of the old way.

Over donuts, I digest these remnant Ideas. I take in their existence, not the memes themselves—I couldn't show you how to tie the Crow's "special tie," but I do know that it existed, that it was special, and if need be, I have a book that holds the information should I wish to learn it. Not that, I decide, *The Indian Tipi* holds the memes themselves. Among the goggle makers there is an ongoing debate about whether memes can exist in books, on audiotapes and video disks, or etched in stone tablets. For my money, they can't. I see memes as *functional* not structural units. When out of a mind, they no longer exist, and where they did exist, there is only lifeless, dry information. The same is true for genes. When in a cell, they can function. They work with enzymes. They switch on and off. They manufacture proteins in the cell water. They exist. But when out of a cell, they no longer exist; you're left with only DNA, fancy, information-rich dust that will remain inert until plunged back into another cell—at which point, like sea monkeys thrown into an aquarium, the genes pop back to life. My guess is that the same is true for memes. *The Indian Tipi* is not "memetic." It needs a hungry mind to become memetic. Until the book is opened and read, it's just info dust: inert, dormant.

Sixty-odd years ago, when Reginald was writing *The Indian Tipi*, the memes of the Crow's special tie Idea and all those other fragments of the old way were very much alive in *his* mind. He'd sought them out, found primary sources and extracted them. He'd dedicated his life to collecting them, testing them, and then finding ways to pass them on to as many other minds as possible. He and Gladys would run courses on tepee construction from their home in Wyoming. They would tour the Indian fairs to teach the Indians their own old dances, exactly as they were supposed to be danced. They made a rig so that they could carry their tepee on top of their car, and traveled about the nation, to museum curators and historians, to country shows, putting their tepee up and taking it down, time and again, just so that other minds would hold some of those old way memes. But, still, their need to pass on those precious Ideas was not satisfied. In the middle of the 1950s they sat down to write a book. In a book, they must have decided, they could store those Ideas for the minds they couldn't hope to reach physically: minds in far-off places, minds not yet created. And at some point during the book's writing, Reginald must have come to commit to paper the Crow's special tie. How should he do it? It was a psychomotor Idea, so it was as difficult to pass on as my mum's shoelace tie. Ideally, he would have bred the Idea by demonstration—and had video equipment been widely available, then no doubt he would have—but he wasn't making a film; he was writing a book. So he decided to tackle the problem with a three-pronged attack. First, he drew a careful illustration of the tie. He used a convention for relaying the information of knot making by exaggerating the position of the rope as it twisted and turned about the poles in order to make it as clear as possible. Second, he augmented this drawing with labels and notes, with arrows, with letters, and accompanying descriptions, showing where *a* and *b* should go or be. Third, he wrote a lengthy textual description with extra pointers in the main body of the book. He did all of this because his instinct was that if he couldn't convey the Idea accurately to the interested mind of the future, then it wasn't worth his efforts. In fact, more than that, his instinct was that if he couldn't convey the Idea accurately, he was doing the Crow a greater disservice than if he never bothered at all.

Why did Reginald feel that it was so important to relay the Idea accurately? Why did he go to such trouble? In fact, why did he bother writing *The Indian Tipi* at all? Through these goggles, the meme's-eye view, the answer is obvious. The only parties that genuinely benefit from a dedicated couple who spend all their energies and time in an effort to preserve and spread the right Ideas are the memes that constitute those Ideas. Just as salmon and bees devote their lives to their inherited genes, Reginald and Gladys Laubin devoted their lives to their inherited memes—and to the exclusion of their own genes: The Laubins never had any children. Now, it's insensitive to draw any conclusions—I don't know why they didn't have children, and it's none of my business—but it is not inconceivable that two people with such a strong instinct to collect and disseminate the Ideas they valued so highly would decide to forego children in order to concentrate on that task. And the fact that that scenario is not inconceivable *is* inconceivable unless you place memes at the wheel of our evolution. No creature built and run by genes would ever have the power to choose "dual income, no kids" unless it fell prey to a second replicator, one that took over the project management of that creature's evolution. This regime change in replicators would have had to be carefully orchestrated—it's no good building a creature that entirely rejects its own genes—for that would be a creature that existed on the planet for only one generation, and there's no future in that. But if the second replicator could guide the first to build and run a creature that was primarily interested in *its* survival and reproduction, and only secondarily interested in the survival and reproduction of its genes, then that would be a creature to strive for. It may take a long line of prototypes, but *that* would be the ultimate goal of this peculiar, unique, double-replicator evolution.

And here it is . . . I mean here I *am*, reading *The Indian Tipi*, sitting in the sun, eating donuts and drinking coffee, jotting down notes—the ultimate meme-gene machine, in ultimate meme-gene heaven. Why am I doing all these things? I'm eating donuts because my primate body is addicted to high-calorie foods. I'm drinking coffee because my primate brain is addicted to caffeine. But I'm *continuing* to eat donuts and *continuing* to drink coffee—even though Adam is sitting in the car at the entrance to the campsite beeping the horn, keen to get back on the road—because my mind is addicted to Ideas. They take up all my energies and time.

Why did Reginald and Gladys write *The Indian Tipi?* Reginald, Gladys, and I all know the answer to that: it was so that a creature like me could read it.

On the Origin of (These) Tepees

We head north along the fabulous flank of Glacier National Park, toward the Canadian border, and with *The Indian Tipi* digested and all the Ideas joined up, Ads (at the wheel) is a captive audience to my summation . . .

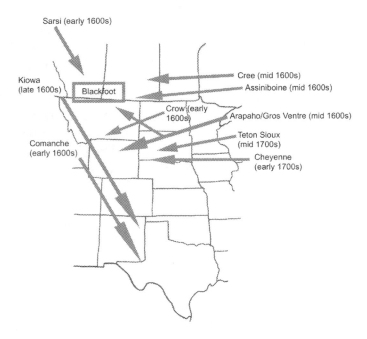

Sarsi (early 1600s)

Kiowa
(late 1600s)

Blackfoot

Cree (mid 1600s)

Assiniboine (mid 1600s)

Crow (early
1600s)

Arapaho/Gros Ventre (mid 1600s)

Comanche
(early 1600s)

Teton Sioux
(mid 1700s)

Cheyenne
(early 1700s)

The approximate dates of Plains Indian migration onto the plains.

JH: Ads, we happily accept that the Plains Indian was a noble ancient, but it turns out that this isn't the case. I knew that the Sioux was a recent arrival on the plains, a domino that fell west when the Chippewa domino hit it, but what I now know, thanks to the Laubins, is that almost all the tribes that were in position on the shortgrass when the Europeans joined them in the early 1800s were relatively recent arrivals to the plains; they'd only just picked up the "old way"—

AH: Dominoes?

JH: Yes, for example, the Sioux displaced the Cheyenne in the early 1800s from the Black Hills, but the Cheyenne themselves had came onto the plains only a century before that, from the same eastern woodlands. They had only just secured the Black Hills from the Kiowa when the Sioux kicked them out—

AH: The Kiowa?

JH: Yes, and the Kiowa may have been safely ensconced in the far south of the plains, looking like Plains Indians, when the Americans got there, but only decades before, they had completed a century-long migration from just south of here, western Montana. They'd come the other way, over the Rockies from the West, taken up the buffalo hunt and the tepee like the others, and, finding no vacant spot to dawdle, drove south and east until they eventually settled with the Comanche in the southern plains—

AH: The Comanche?

JH: Yes, and the reason that the Kiowa could find no spare space up here was that in the century before that, the 1600s, the Cree, Assiniboine, and Arapaho had all arrived, independently, on the *northern* plains—

AH: The Arapaho?

JH: Yes, and almost as soon as *they* got here, they split, with one band, which was to become the Gros Ventre, venturing up toward the Canadian border, and the other, the "Arapaho" themselves, journeying eight hundred miles southwest into Colorado, where they were later to link up with the arriving Cheyenne to take on the Comanche—

AH: The Comanche? [after short pause] Oh, I've already said that.

JH: —*who* were also immigrants to the plains. They'd traveled down at lightning speed in the 1600s from the extreme west of Wyoming, right up against the mountains to dominate the south—

AH: So *when* are we now?

JH: Early 1600s.

AH: Is that it?

JH: Well, that's about it for the dominoes, because the European colonists weren't making their presence known much before that. However, tribes were still migrating. The three tribes of Knife River—the Hidatsa, Mandan, and Arikara—they arrived on the eastern edge of the plains five hundred to six hundred years ago. Sometime after that, the Crow broke off from the Hidatsa and moved into Montana to become proper tepee dwellers—

AH: And what about the Blackfoot?

JH: Exactly. Of all the tribes we know now, the Blackfoot are about the only people with a history on the plains of over five hundred years. That's why they were called "the kings of the plains." Blackfoot originally lived to the east of here—they were pushed up against the mountains when the Cree came—but they were plains people even before that.

AH: So the Blackfoot invented the tepee?

JH: Well, the history of this region is sketchy and complicated, but I'm certain that the four-pole design was first. After all, all the earliest plains tribes: the Blackfoot, the Crow—I haven't mentioned the Sarsi, but they probably joined the Blackfoot in the North about the same time as the Crow—and, the Comanche in the south: they all had four-pole tepees. So either the four-pole came first or they all swapped from a three-pole tepee for some reason, and I can't imagine that, because the Laubins say time and again that the three-pole tepee is a better design. It's just as Kenny the Ute said, it's easier to put up and more stable when up. The Laubins suggest that any tribe living in the windy open plains of the east ought to use a three-pole for that reason, so I can imagine a three-pole evolving from a four-pole, but not the other way around.

AH: So who invented the three-pole?

JH: I think it was the Mandan, and here's why: There's another four-pole tribe that I've not yet mentioned, the Shoshone. The Shoshone used to be a plains tribe. They lived virtually alone on the plains with the Blackfoot from sometime distant until about the 1600s, when they left the plains and moved westward over the mountains and into the Great Basin, presumably in response to the arrival of the other tribes from the East. Now, the Shoshone four-pole tepee is odd. First of all, it's got smoke flap pockets, which my guess is they *did* invent; and second, its front two foundation poles are on either side of the door, at five thirty and six thirty—so close together that it stands like a three-pole tepee. In fact, the first time the Laubins saw a Shoshone tepee they thought it *was* a three-pole tepee.

So, starting with a Shoshone design, all you'd have to do to "invent" [I perform air quotes] a standard-issue three-pole tepee is to omit one of the front two poles when putting it up. That could have happened by accident or on purpose, but I don't think it happened until the early 1600s. The Comanche are a branch of the Shoshone that migrated south at that time, and they took with them the same strange four-pole tepee design, so it must have been after that. But *after* that, *every* tribe that came on to the plains adopted the three-pole with smoke flap pockets. Who do I think invented the three-pole? I think it was the Mandan. They were the earliest of the three-pole tribes. We know that they were in contact with the Shoshone, because they stole their women—remember Sacagawea, the squaw who went with Lewis and Clark from the Mandan village? She was a stolen Shoshone woman. So what if they got the tepee Idea from the Shoshone and modified it, by accident or on purpose, because they needed a more stable tepee out there on the eastern plains, in the open, hunting their buffalo? Because the Mandan traded with everyone, the three-pole design could easily have spread to all those new tribes from there.

But what's fascinating is that even though the three-pole tepee design could be less than four hundred years old, every one of those new tribes built one that looked slightly different. Isolated on their islands, the tepee Ideas diverged, slightly, even over that short period of time—the smoke flaps were different shapes or the smoke flap pockets were different; the Assiniboine tepee went through a strange mutation: they started stacking the top poles in the opposite direction—but none could be mistaken by an expert. They were all different species, and I suspect that on top of random memetic drift, the tribes liked it that way. Everyone likes to be unique.

AH: So you've solved it?

JH: No, I've not discovered the origin of tepees, I've just got a theory about the origin of *these* tepees.

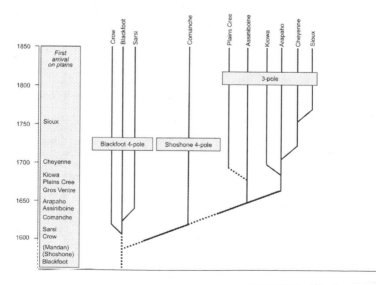

My guess at the tree of tepee descent.

I show Ads a tepee tree that I've sketched on the laptop. He can't really look, because he's driving, but he nods appreciatively all the same.

"But what I don't understand," he says, "is why they all piled onto the plains so recently. I get that the Europeans made them move, but were the plains empty before then except for the Blackfoot and Shoshone? Why weren't more tribes hunting buffalo on the plains before the 1600s?"

The arrival of the Europeans forced tribes onto the plains for two reasons: it *pushed* tribes onto the plains from the East with the domino effect, but it also *pulled* tribes onto the plains by accidentally delivering to the native population a new buffalo-hunting technology, the domestic horse. The horse made Plains Indian life an order of magnitude more productive and less painful. Before horses, tribes on the plains were severely restricted in their capacity to hunt buffalo in large numbers. They were also severely restricted in their mobility: they couldn't follow the herds as they would later do. Before the horse, the only way of moving a tepee over the plains was to drag it yourself or to tie it to a dog. "The dog days," the Plains Indians called them. The phrase conveys a tough, bitter life, a life that you wouldn't choose. And few did. The Blackfoot and the Shoshone were possibly the only tribes living permanently on the plains in those days, living in much lower densities than they would come to do before the end.

From the 1600s, horses were robbed from the Spanish in the South, bred by the Apache and Shoshone, and ultimately, through trade and theft, disseminated throughout the plains. The horses rescued the incoming outcast tribes by providing a means for them to prosper in the most unlikely of places, the Great Plains. Who knows how long the buffalo herds would have withstood this new hunting technology. In the end, the "iron horse" cut short that experiment.

AH: So, what now?
JH: We go over the border, into Canada. Up there are relics from those dog days.

PART V

Mysteries Solved

14

The Past

The Dawn of the Smelly Heads

"Wake up you sleepy head" . . . gambrel roof . . . sagebrush . . . "It's not your fault, Jon" . . . *"Look out my window and what do I see"* . . . weird fencing, not seen that style before . . . "I know, Ads" . . . coffee smell . . . what time is it at home? . . . "No, listen to me. It's not your fault" . . . eighteen-wheeler. Mack? . . . I need a wee . . . "Don't mess with me, Ads, not you, not you, Ads" . . . itch, breathe, blink . . . "Canada, 8 Miles" . . . *". . . a puzzled man who questioned what we were here for"* . . . slow down! . . . push foot . . . "Pronghorn" . . . "Where?"

Within my skull, juggling my life, forming new memories, flooding the world with context, requisitioning breaths, heartbeats, and eye blinks, indulging Adam's sense of humor, looking for onward directions, listening to Bowie and calculating our relative position as we fly through space at seventy miles per hour is the most complex object in the known universe, the human brain. No other animal has anything like it. It contains over one hundred billion brain cells and one hundred trillion brain cell connections. It is a marvel, and I'm extraordinarily lucky to have one. The question is why *should* I have one?

If you remember, that was mystery number one, "the mystery of the past": how come we humans evolved a device that we don't appear to have needed. If the survival and reproduction of our genes was the goal, then why were we—creatures that, on the face of it, could survive and reproduce with only a few minor tweaks to our design—subject to a roller-coaster bout of evolution that lasted from around four million years ago to around two hundred thousand years ago? What pressures were working on our genes during those years of heady "progress"? And why did our brain evolution then screech to a halt two hundred thousand years ago?

Mainstream science doesn't know the answer to these questions. To date, the only partial explanations have come from a gang of wily hard thinkers called the evolutionary psychologists. These people have spent the last two decades reaping the benefits of the application of the gene's-eye view to the study of human behavior. They've found the selfish gene at work in a plethora of our activities, from our adultery to our child abuse to our bitter office politics to our lies and deceit. It's not always pleasant reading; it often makes us feel as if we're little more than apes in jeans. But the evolutionary psychologists tell us that it's worse than that: we're actually little more than *genes* in jeans. And that is the point at which the general public becomes skeptical.

Yes, we say, we can see that our ape selves will have a weakness for donuts and coffee, that our genes will have little love for a stepchild, and that it would make sense, from her genes' point of view, for a woman to marry a devoted, if weak husband while bearing the love children of an extramarital "bastard." How do you explain, however, that we're not all devouring donuts and coffee 24/7, that stepchild abuse is extremely rare, and that so many women end up in a dead-end relationship with a selfish prat? In most instances, in most of our lives, the gene's-eye view doesn't work! Unlike every other species on this planet, every other product of natural selection, genetic fitness is not the primary mover in our lives. So, what is?

That's where the meme's-eye view comes in. Through these goggles, the reason why Ads and I are so uniquely weird is simple: we're the products of a uniquely weird evolutionary project, the most remarkable that Life has ever undertaken, a project driven by not one but two replicators. The gene does still get something of a look-in—which is why evolutionary psychologists can explain some of our behaviors—but running the Humanity Show these days, and thoroughly enjoying the experience, is the meme.

Why did the gene relinquish its power? How did the meme take the wheel, and what did it do next? That story is the one that solves the mystery of our past; and it starts in the age of the dinosaur.

The first mammals evolved approximately two hundred million years ago, at a time when dinosaurs ruled the world. They had a hard Life. The spectacular success of the dominant dinos meant that if our ancestors' genes were to survive and reproduce, they had to come up with a radical new design. And they did. They pulled off the miraculous: they found, quite by accident, that, even in the Jurassic, half of planet Earth was still virtually free of ravenous dinosaurs, the half that, at any point, was facing away from the sun. Dinosaurs didn't come out at night, so our ancestors did.

But functioning in the dark required some adaptive work. In order to find their way about, the early mammals pumped most of their R-and-D budget into olfaction, the sense of smell. And that had consequences. Olfaction is different from the other senses. Three of the other senses— vision, hearing, and touch—operate with only one or two types of sensory receptor cell. Lying dormant in their sense organs, these cells come to life only when the form of energy they are sensitive to, light energy or mechanical energy, arrives at their particular site. When that happens they send a nerve impulse off to the brain, and the brain uses its knowledge of the relative position of each receptor to build an "internal analogue," a facsimile of the real-world sight, sound, or feeling. When my fingers touch the Chrysler's plastic door handle, I "feel" toughened plastic because my brain's somatosensory center builds a "point-to-point" topographic representation of the firing receptors in my fingertips, and this makes me *feel* that I am feeling toughened plastic. Sights and sounds are perceived in a similar way.

But olfaction is different. The relative position of the sensory receptors up our nostrils is not important because airborne smells don't possess any "physicality"; it's their presence or not that matters, not where in the nose they land. In addition, our noses house not one or two different types of receptor, but hundreds, each a slightly different shape, so that we can detect the presence of hundreds of slightly different-shaped airborne chemicals. When a chemical drifts into our nostrils, it will happen upon the receptor built for it, and at that point, the message "it's here!" is sent at high speed to the olfactory cortex of the brain. But one impulse is not enough. Every odor you recognize is comprised of a cocktail of airborne chemicals, so when, for example, I feel that I'm smelling the latest coffee I've bought myself, sitting in its cup holder, what I am actually doing is detecting dozens of different chemicals in the air with dozens of different receptors in my nose. These all relay their independent signals to the olfactory cortex in my brain—and that's where the magic begins.

The olfactory cortex, in contrast with the orderly point-to-point architecture of other sensory areas, is a complete mess. It doesn't need to be tidy, because smells have no physicality. So its cells fly in all directions, bumping into each other at random. When the dozens of messages that together mean "coffee smell" arrive independently at the cortex, they are each handed on to dozens of other adjacent cells, then on to others, then on to others, each message triggering its own unique chain reaction in the cortex. At some point—a fraction of a second later—there will be a cell in the middle of my olfactory cortex, in all that mess, that receives all the component messages at the same time. And when this happens, I smell coffee. That cell is my "coffee smell" cell, the specific site of my perception.

Computer scientists would call such a circuitry "random access," just like the memory of your computer. Olfaction is a sense fit for the digital age. And because of that, it has some unique properties:

▲ It can accommodate new experience. If you happen upon a new smell, a new combination of dozens of airborne chemicals, it's no problem, somewhere in your jumbled olfactory cortex there will be a cell that exists at the junction of that new combination of impulse firings, and *that* cell will become the site of your new smell perception.

▲ It can be modified by experience. The more a particular site of perception is activated, the more the connections that lead to it are strengthened. This is a simple form of learning.

⚜ It can grow bigger. Random-access circuits have no upper size limit; the larger the mess, the more discriminating and powerful they become.

So the random-access circuit that our ancestors accidentally prioritized had all sorts of applications beyond olfaction. At its most fundamental level, it was a way of identifying a pattern of neuronal firings; it didn't matter what those firings stood for. What if the random-access circuit was extended so that it could collect the secondhand inputs of sights and sounds and feelings, too? What if it got wired up to personal memories and emotions? With enough random-access jumble, you could have a cell that fired only when you smelled coffee *and* donuts *and* saw colorful vinyl *and* eager, young servers behind a counter, *and* the color green *and* the soothing sounds of South American music: you could have a "Starbucks cell," something tied into memories and emotions, something *extra*perceptual, something *conceptual.*

And if you can flag up a conceptual Starbucks, you can, in theory, flag up anything. With enough random-access circuitry, enough connections, you could spot any number of firing patterns. You could identify gambrel roofs and pronghorns and melodies and lines from the film *Good Will Hunting.* You could break the whole world up into "cortical code." You could associate everything with everything else. You could build a supreme model of the universe in your head. You could derive something meaningful from the world about you.

And that's what the mammals began to do. Soon after they became creatures of the night, mutant mammalian brains began to appear that took the random-access circuit of their olfactory cortex and ran with it. The neocortex—a discrete, senseless jumble of neurons—evolved early on in mammalian history. It lay like a sheet on top of the rest of the brain, within touching distance of all the bits and pieces underneath it. In the earliest mammals the neocortex was tiny, but within a relatively short space of time it had trebled in size as only random-access circuits can. And as it grew, the things it could do expanded exponentially. The larger the neocortex, the more rounds of connections its mammal could build on top of its original perceptions, and the more abstract its work could be. Soon, the majority of the mammalian brain's activity was happening in the neocortex. Looped networks developed, allowing impulse chain reactions to be held in the neocortex for a neural age: several seconds.

This enabled the mammals to ditch their innate responses and start to consider and plan before acting—to think about everything. Mammals were becoming very smart.

And then suddenly, sixty-five million years ago, something unexpected happened. The dinosaurs disappeared. There was a new dawn, one that even the timid night creatures could broach. And the meek (but smart) inherited the Earth.

Border Crossings

The border between the United States and Canada has traffic lights. We sit third in line, waiting for green. Our onward journey requires us first to get out of the United States, then drive across no-man's-land, then request admission to Canada. The first of these three steps looks ominous. The American border is fitted with a glamorous new building that smothers the road. A dozen metal outbuildings behind it are guarded by a high fence. As we edge forward we spy young, closely shorn customs officials in dark glasses. It is then that I realize our passports are in the trunk. The green light pings on.

Thousands of years ago, several hundred human beings waited for a different kind of green light at the very northeastern tip of Siberia, although they didn't know it. They were destined to become the first humans to get to the New World. Their crossing was also a three-step process:

⚜ STEP 1: LEAVE SIBERIA. Leaving Siberia was not an option until the planet's temperature dropped by a few degrees, the rivers froze in the mountains, and the oceans froze at the poles. Once this happened, the cold, gray sea that had slopped about to the east of the tribe, limiting their onward expansion, relinquished what lay beneath: the huge, sodden plain of Beringia. The draining of Beringia took a dozen generations, but once it was dry enough to sustain grasses, the tribe, who were buffalo hunters, were inevitably drawn on to the new flats by the vast herds of buffalo whose brains could see only as far as the fresh shoots immediately ahead. The hunters had to stay close to their herds. In Beringia the nights were always below freezing, the winds were treacherous, and the ground thawed only in high summer.

Without the herds' pelts and dung, the hunters would never have survived.

⚜ STEP 2: BECOME MAROONED IN NO-MAN'S-LAND. The next step was even grimmer. A hundred generations later, the tribe was still camped in Beringia, up against the crags that today define the westernmost coast of Alaska. They were still living off the buffalo herds when the temperature, which had been climbing unnoticed for two centuries, reached a critical level. The ice melted in the poles and in the mountains, and the cold, gray sea returned to separate the Old and New Worlds once more. At first no one noticed—the sea gradually welled inland over the horizon to the west—but as the land became subsumed by greedier tides, the tribe recognized that they could either remain in the east of Beringia or return to their homelands in the west. In the end the choice wasn't theirs. The majority of the buffalo brains chose the east, so that's where the majority of the tribe remained, and before they knew it, the north and south fingers of the sea touched each other, and the buffalo hunters were isolated from the rest of humankind. To their north, west, and south, the ocean raged again. To their east, the land rose slowly up to a mile-thick wall of ice. There they remained, stuck between the ice and the sea, red lights in all directions, for thousands of years.

⚜ STEP 3: ENTER AMERICA. Ultimately, a green light came on: the same rise in global temperatures that had blocked their path back west relented and delivered a way out to the east. The first thing the tribe noticed was that the land all about them began to flood—not with seawater this time, but with freshwater. The buffalo fled from the lowlands in response, driving upward and eastward to escape the growing coastal marshes. The tribe had no option but to follow them, and when they did, they came upon a changed landscape. The legendary mountain of ice that blocked their path to the east was dripping in the sun, and in the very middle, between the mountains, a corridor had been carved out of the ice. It wound eastward through the ranges, a tunnel in the ice sheet, its floor carpeted with fresh grass. The buffalo unthinkingly entered it, so the tribe did, too. For fifteen hundred miles they wandered down the corridor in the wake of the herd, ice escarpments towering above them on each side, until they emerged onto an endless high plain covered in golden grass. The herd, in its excitement, trotted out onto the Promised Land, followed by the tribe. Although they didn't know it, both the prey and their predator had just gained sole ownership of an ice-free, human-free territory twelve million square miles in extent.

The dates of this three-stage border crossing are hotly disputed. Glaciologists, archaeologists, human geneticists, and the descendents of the tribe themselves tend to disagree, but dates aside, that's the consensus on how America was first peopled. And it's how America first got *Bison bison*, the buffalo. There was another species of *Bison* living in America when the herds and their hunters first entered the Great Plains. The native American buffalo, *Bison antiquus*, was much larger in stature, slower to breed, and heavier on its feet—all features that evolved to combat the attacks of the sabertooth cats that had historically preyed upon it. But they were characteristics exactly unsuited to combat the attacks of their new predator, the human being, and within a few millennia they'd all been eaten.

The only other accomplished plains herbivore in the Americas at that time was . . . the horse! Ancestral horses had jumped the Bering Strait traffic lights several times in the previous three million years, skipping back and forth between Old and New World in the required three steps as the ice waxed and waned. Those horses that found themselves in the Old World had to adapt to the emerging hunting strategies of humans—and because they adapted biologically as the humans' hunting strategies adapted culturally, they kept in step. However, the sudden arrival of skilled hunting humans armed with "atlatls," or spear-throwers, in North America was too much for the New World delegate of the horse taxon to cope with, and it took only a few thousand years for them, too, to be eaten. So it was that the invasive *Bison bison*, which had adopted its small size, quick reproduction, and fleet-footedness precisely to avoid extinction at the hand of the human harvesters over tens of thousands of years, spread out gaily across the continent to eat all the grass. Their hunters did likewise, to eat them.

So the question that I posed to Ads back in the Dakotas—which is an authentic Great Plains animal, the horse or the bison?—was not as easy to answer as he thought. The horse lived in North America for millions of years before the bison that we know arrived. And that bison has been here only as long as we have. Indeed, they are the reason we first came here. They led us here.

We pull up alongside the young U.S. customs officer. "Passports," he says without a smile. "I've just realized they're in the boot," I say. "The what?" he says, and Ads opens his door and steps out of the car to go get them. Immediately the officer skips back from my window, pops his holster, and stands in action pose, hand on the butt of his revolver.

"Get back in the car, sir!" he yells, and a colleague immediately rushes over to take up an alternative shooting angle so that they can kill us better.

"Okay," says Ads, "sorry."

We are told to reverse to the end of the line and get our passports out at a safe distance. We apologize again and again, admit our stupidity, deprecate ourselves as only the English can, until it pains them.

"Sorry won't be good enough next time," they spit.

Twenty minutes later we arrive at Canada's border. "Hi, guys," says a man with laugh lines. "Whatcha doin'?" We tell him our plans, he assures us that we're going to have a great time, and we believe him. We trundle off with an "all the best" into an identical countryside, but the cultural landscape has changed entirely. The speed limit is now one hundred, because it's in kilometers per hour. There are recycling bins all over the place. The horizon is littered with wind farms. The barns are odd: gambrel-roofed, but with wide hips. The houses look like English houses, with gray pebble-dash walls and dormer windows. The markings on the road are different, and because there are lines across the right of way, I find myself repeatedly slamming on the brakes. But the Canadians on the road with me don't mind. I have Texas plates, and they still don't mind.

Food for Thought

Armed with a brain of expanding random-access circuitry, the mammals that inherited the vacant world at the end of the age of the dinosaurs took to the trees, the deserts, the coasts, the rivers, and the grasslands like ducks to water. They invaded every sizeable vacant niche from deep-sea squid eater (sperm whale) to desert tuber eater (naked mole rat). Their design—various quantities of hair, warm-bloodedness, milk and placentas, and, most important, a neocortex—evolved to suit the opportunities. Quite by accident, the mammals found themselves to be an order of magnitude higher up in design space than anything else on Earth.

The group that we are most attached to, the primates, took up the vacancy of tree hugger. The first of the group literally did just that, straddling a trunk for most of their lives, but evolution sought to improve their arboreal dexterity. Soon they had opposable thumbs; long arms; long legs and long, balancing tails; and forward-facing eyes, so that they could accurately judge the leaping distance to the next tree. With this sort of equipment, the tree huggers soon had no reason to descend to the ground. Everything they needed was suspended above the Earth, so they spent their whole lives halfway to the sky.

The first of the group, the true tree huggers, were insect hunters. Insects lived everywhere in the forest, so these lightweight, bug-eyed primates didn't have to worry too much about territories. They lived alone on the creepy crawlies that surrounded them, partnering with another of their kind only when the urge to procreate took hold. But with time, the primates extended their ambitions. They ventured into leaf eating.

Leaf eating is not as easy as it may sound. Forest trees pump toxins into their leaves as they mature, so to engage full time in leaf eating, you must either develop specific antitoxins and stick to the trees whose leaves you can digest—which is fine for beetles, but not for animals as large as a primate—or search the canopy for only the youngest, mildest leaves. Now a territory becomes important, because your rival primates are after exactly the same baby leaves as you. The leaf-eating primates had to get organized. They teamed up into closely related groups and began to defend the prime real estate. Group living—"sociality," a strategy that would turn out to be an extremely significant tooth on the ratchet— brings its own necessities. You need to be able to recognize everyone in the group. You need to be able to communicate effectively with your group members. You need to be able to trust each other when conflict with another group becomes likely. So the leaf eaters began to get better at these tasks. They evolved signature smells and visual cues to assist them in recognition. They developed a suite of vocalizations—barks and chatters—to signal the arrival of enemy predators or rival bands of primates, or just to announce a reassuring "I'm still here" in a world in which you could barely see more than twenty feet in front of you. They began to groom one another rigorously, initially to extract parasites, but ultimately to cement the social cohesion that a group of primates needs to function.

This interindividual work took brains—or, more specifically, a neocortex. The smells, sights, sounds, and feelings of sociality would work only if the neocortex sequestered them and imbued them with association, context, meaning. Natural selection thus forced the neocortex to enlarge so that it could keep up with the growing demands of social networking. The larger the neocortex became, the larger the group could become, the larger their territory could become, the greater the mass of available, edible leaves would become, the more offspring could become, the larger the group would become, the greater the force on the size of the neocortex became. Fed by this cycle of progress, the neocortex of primates ballooned, especially when they took the next step and changed their diet to fruit.

In an evergreen forest, there are always trees in fruit. The trick is finding them and then securing them from rivals. To do both, you need yet bigger social groups, and as we are now well aware, this means yet bigger neocortexes. So it was that, barely twenty million years after the dinosaurs disappeared, our direct ancestors found themselves dominating the daytime world of the tropical rainforests with their enormous, complex social groups and their enormous, complex neocortexes. They would sit around in the heat of the day eating fruit, grooming one another, playing to exercise their enormous brains, kings of their domain, with all the time in the world. It was at some point after this that they discovered how to ape.

The Art of Aping

Highly social, brainy primates with time on their hands are able to watch the actions of others and copy them, just as I watched my mum tying my shoelace and then copied her. Biologists call this skill imitation. Anyone else might call it aping.* As I've mentioned before, aping takes a lot of mental prowess. It's not easy to analyze the movements of others, distinguish which is causative and which is superfluous, and then repeat those causative actions, muscle by muscle, with your own body.

---

*Or is it "parroting"? If you were to press a person on the street as to what the difference is between parroting and aping they might answer that parroting is copying *without* thinking, while aping is copying *with* thinking. If so, that's pretty much how biologists see it, too.

But one can easily understand why the genes of our primate ancestors would have encouraged such exhaustive brain work: copying the useful behaviors of others will always be useful to a primate, and hence to the genes *in* the primate. Learning from others how to use a stick to extract termites from a termite mound, for example, adds an easy, dependable supply of vital proteins to a primate's diet. And a primate that never goes short of protein will be stronger, more alert, and more likely to defeat bouts of sickness. It will, in all likelihood, have more offspring, so inevitably the genes that gave the primate its imitating skill in the first place will, over time, become more common in the gene pools of subsequent generations. Good imitators are also valuable to the genes of their mates, because, in the bodies of their mutual offspring, their mates' genes will become mixed up with the desirable imitation genes. So it's also in the interests of the genes of others to make *their* primates sexually attracted to the best imitators. As psychologist Susan Blackmore, who first outlined this theory of evolution, might say: imitators became first successful, then sexy—it sounds familiar, doesn't it?—and that's how the cerebral technology of imitation spread.

Presuming that our primate ancestors were gifted in the same way, it now becomes possible to appreciate how true culture took off and then took over, at some point, on some page of our ancestry flip-book. The art of aping enabled primitive psychomotor behaviors—such as using a termite stick—to pass among the group, infecting one mind after another. This capacity for culture was so advantageous that natural selection promoted the genes that coded for the imitation technology. By tying imitative ability to sexuality, natural selection forced the coming generations of primate brains to develop it further. A "gene rush" must have ensued—a rampant bout of neural R and D focused entirely upon building more and more proficient imitative brains. And this meant building brains that were not prewired, but that were open to change and able to learn; plastic brains that could build themselves as their lifetime unfolded; brains that were less and less determined by the genes.

Desperate to get a taste of the successes of the sexy imitators, our ancestors' genes accidentally/automatically relinquished their control over the structure and function of the most important organ in the body. This may not have mattered had the advent of imitation technology not done something unique in the history of Life on Earth. It brought into being a new system of inheritance: culture, a brand-new mechanism by which information could be handed from one individual to another *without genes*. So all of a sudden there was a vacancy, a space for a new replicator.

My fellow goggle wearers suggest that this new replicator, the meme, flickered into life in the minds of our ancestors at some point in the last six million years, when our ancestral line split with that of the other apes. Once formed, it replicated and mutated and adapted and, leaping from brain to brain, automatically/accidentally brought about all of the Ideas of human culture.

That would, indeed, explain everything. So why aren't all biologists talking about memes? Why are these "meme goggles" so rarely worn? The chief objection in the mind of meme skeptics remains: If these new replicators are found everywhere and throughout our brains, then what does one look like? If memes are so prevalent in the human mind, then "show me one."

The fact is, we can't. We don't yet know what a meme looks like. That doesn't necessarily mean that we should dismiss their existence—the existence of Mendel's genes was mere speculation for a century before Watson and Crick "saw" them for the first time with an X-ray diffractor*—but it certainly presents a barrier to widespread acceptance. However, it could be that the first pictures of memes are coming into view.

In 1996, and quite by accident, researchers at Giacomo Rizzolatti's laboratory in the University of Parma, Italy, got their first glimpse of the science behind the art of aping. They were studying the neuronal activity of the premotor cortex of Rhesus monkeys. The premotor cortex is known to select, plan, and order the execution of actions. The team was trying to discover how monkeys use it to elicit their hand movements. Electrodes sited in the monkeys' brains were wired up to a speaker in the lab such that each time a monkey thought about moving its hand—click!—a sound would come from the speaker. Work was going well; the lab had heard clicks each time the monkeys reached for offerings of fruit. Then, one day, when they ran out of fruit, the team took a tea break. Halfway through this tea break, the speaker suddenly clicked. Rizzolatti glanced over at the monkeys, expecting to see one of them reaching for a newly spied grape—but no, the monkeys were calm, seated, still, and silent. In fact, the only thing they were doing was reversing the experiment and watching the researchers reach for *their* food. Rizzolatti looked

*The machine they used to take the first pictures of DNA.

back at his team. One of them had a banana in his hand. "Do that again," he tentatively said. The researcher reached for another piece of fruit. Click! There it was again. And again, and again: every time one of the lab technicians reached for a piece of fruit, the neurons in the monkeys' brains that organized exactly that action fired. The monkeys' brains were going through the motions of reaching for fruit whenever they watched someone else reach for fruit.

The experiments have been repeated and developed, and we are now beginning to learn a lot about these so-called "mirror neurons," which fire in primate brains when they observe another primate behaving. For example, we now know that they are active all over the brain, not just in the premotor cortex. We know that the perceived intention of the observed action is important; different mirror neurons will fire if the monkey imagines that the researcher is reaching for fruit not to eat but to put back in his lunch box. We know that there are mirror neurons associated specifically with mouth movements, and that, again, you find different patterns of activity in these neurons depending on whether the mouth is moving while eating or moving while vocalizing. So monkeys are apparently able to observe another's movements closely, make assumptions about the intentions of the movements from surrounding evidence, and replicate the firing patterns of those movements in their own brains. All of which means that primates armed with this involuntary brain technology are more than just social; whenever they watch one another, their nervous systems are, in a practical sense, physically connected.

This sounds like the sort of hardware that may have been hijacked by a mischievous replicator looking to transfer from one brain to another. Sure enough, in 2010 the first concrete evidence of the existence of mirror neurons in humans was published.[1] Undergoing exploratory work before surgery, twenty-one epilepsy patients volunteered their implanted electrodes to the mirror neuron researchers. These electrodes were not placed in the premotor cortex as they were in Rizzolatti's monkeys. Nevertheless, when the patients performed and then watched others perform facial expressions and hand movements, researchers discovered that the same neurons fired in both events. The only difference was that the fireworks were more subdued when the patients were merely watching the activity.

So here's one possible sketch of the meme: perhaps a meme is a particular firing of a particular pattern of neurons in our huge random-access brains. Perhaps an Idea is the unique fireworks display created when hundreds, if not thousands of these memes light at the same time in our neocortex.* And perhaps we are the crème de la crème of Dennettians, the best thought-swappers, the most artful of apes, because we are exceptionally talented at mirroring these unique fireworks displays in our own minds every time we perceive them in the minds of others. In sum, perhaps the involuntary twitches of our copycat mirror neurons could be the very essence of our capacity for culture.

But we should qualify what would be required for this mirror system to work. For an Idea to transfer successfully from one mind to another, a pattern of neuronal activity in the donor brain must be reproduced in the brain of the receiver, but this duplicated fireworks display does not need to be *structurally* identical to the fireworks display in the donor brain. Indeed, since no two human brains *are* structurally identical, that must be impossible. It does, however, need to be *functionally* identical, or at least functionally *near*-identical (because there's nothing wrong with a little mutation). In other words, the reproduced meme must site itself in that mass of connections in such a way that it triggers an almost identical array of other memes whenever it is lit. The value, the "meaning," of any Idea is only ever a function of the other Ideas it is associated with. If the art of aping can light functionally identical fireworks displays in the minds of others, then undoubtedly a form of replication has occurred, and natural selection can get to work.

The meme may be coming into view. But, excited as I am about research into mirror neurons, I don't need to wait until I see a picture to believe in memes. Through these goggles, the presence of memes is obvious, never more so than when you stare into the eyes of our cousins, the chimpanzees. They, too, share the art of aping and the power of culture (primitive though it may be). They even share the vast majority of our genes. (The difference in genes between you or me and a chimpanzee is only ten times greater than the difference between you *and* me.) What makes you or me different from a chimpanzee is not these things;

---

*And if so, the traditional imagery of an Idea as an illuminated light bulb is pertinent—light bulbs are solid and inert; they emit light only when the arrival of electricity alters their physical properties—the light is a transitory dynamic phenomenon. So it is with Ideas. The neurons are always there. It is only when electricity passes through them in a certain way that they come to life.

it is, I believe, that due to some quirk of circumstance, the memes never took hold of the evolution of the chimpanzee. In our case, they did. With its parent replicator doting on its every whim, the bastard son, the meme, took the wheel and built itself the ultimate meme machine. And if you want to see what *that* looks like, just glance in the mirror.

Head-Smashed-In

Barely an hour and a half north of the border, we spy a break in the plains where a pronounced escarpment rips across the landscape. We veer toward it, and a stacked, camel-colored crag comes into focus, flat as a tabletop, rising out of the silvery green grass sea like the flank of a supertanker.

In 1792 a British mapmaker called Peter Fidler was sketching the very same crag while conducting a survey trip for the Hudson Bay Company. Near its base he could see a village of the Blackfoot tribe. Despite the fact that Fidler was the first European they'd ever met, the Indians already had plenty of horses. Clearly the horse trade that began with Apaches stealing from the Spanish almost two centuries earlier had infiltrated all parts of the plains by the late eighteenth century and ended the dog days forever. Nevertheless, Fidler's stay with the Blackfoot revealed that they were still getting used to their new "elk dogs."

The tribe had taken up position below the escarpment because they were planning to carry out a dog days–style buffalo hunt. Before the arrival of horses, the only way to kill a large number of buffalo on the open plains had been to run them off a cliff. The camel-colored crag Fidler had just added to his map was one of the best buffalo hunt cliffs in the northern plains, and hence a priceless piece of real estate to the tribe. The Blackfoot referred to it as the Head-Smashed-In buffalo jump, after an unfortunate incident in which a young boy made the mistake of watching the hunt from below the cliff, a dramatic but deadly vantage point. The tribe was friendly toward Fidler (in these days before exploitation) and requested that he join them in the hunt.

As it turned out, it was an extraordinary stroke of fortune for historians that he did. Fidler got a rare glimpse of the dog days as he joined the Blackfoot in their preparations. Scouts had pinpointed a herd to the west of the cliff, nestled in the Porcupine Hills, where they could be rounded up and ushered toward the jump. It was late fall, the perfect time for the hunt: the cows were fat and hairy, the calves were large, and the bulls had long gone, which meant that the herd wasn't so edgy. The coming cold weather would keep the meat good, and the rest could be made into pemmican for the winter.

The Blackfoot set to work. They first conducted an "iniskim" ceremony, which involved singing to the buffalo, burning sweet grass, and visiting the sweat lodge to shed their human scent. Next, teams went up on to the hills and built the "drive" down which the herd would run. They marked the sides of the drive with spaced cairns of dung, turf, and branches. They dragged buffalo skins along the ground in the middle of the drive, to give it a familiar scent and encourage the brain of the herd in the right direction. They set up a processing camp at the base of the cliff, with ready piles of dung for fuel, skin-lined pits, stones to heat, and water to boil.

Then, the night before the hunt, they told Fidler the plan:

1. A team of young men would work as "buffalo runners." Some would dress in wolfskins and, creeping low to the ground, come up behind the herd, visible against the ridges, to encourage the brain to lead the herd into a "gathering basin," just in front of the drive. Others would dress in buffalo calfskin and crawl to the front of the herd, at the start of the drive, where they would lie in the grass and wait.

2. The rest of the hunters would get into position behind the cairns of the drive, buffalo hides in their hands.

3. When the time was right, the wolves would start to trot toward the herd. The boys in the calfskins would suddenly jump up and bleat like startled young buffalo. Pushed by the wolves and pulled by the troubled cries of calves, the brain would lead the herd into the head of the drive.

4. Once the majority of the herd was in the drive, the wolves would start to run, the calves would start to run, and the hunters behind the cairns would leap out of hiding, flapping their hides violently and screaming.

5. The herd would hit panic speed, a stampede. The calfskin-clad boys would have to run to safety between the cairns as fast as they could, timing their run perfectly so that the momentum of the herd was not interrupted.

6. At the end of the drive, the herd, at full speed, would meet the top of the jump. Head-Smashed-In was one of the best jumps in the northern plains because its topography of small hills on either side of the drive funneled herds to one tight spot, the thirty-foot drop beyond was hidden by a lip at the top of the cliff, and the prevailing wind came from the west, so that the buffalo could not smell the camp below.

7. Over they'd go, in their hundreds, if the jump was expertly performed. Most would be killed on impact, but those still breathing would be quickly dispatched with spears and hammers.

8. The tongues and offal would be eaten almost straightaway. Stomachs would be cleaned out for use as watertight containers. The meat would be stripped and dried or pummeled into pemmican. The bones would be converted into tools or thrown into the skin-lined pits of water brought to a boil by hot stones. Once brittle enough, they would be split and their marrow boiled to make grease and glue. The sinew would be used to make thread for sewing. The hides would be stretched on the ground, pinned down, and scraped clean to make clothing, parfleches, or tepee covers.

This was the plan relayed to Fidler in 1792, but it didn't happen. After the explanation of the traditional hunt, some of the young Blackfoot decided that they should incorporate their horses into the hunt for the first time. Surely the combination of horse and drive would guarantee the most successful hunt ever. Nice idea, but it didn't work. The horses scared the herd before it got onto the drive, then panicked when it stampeded. In the end, only one buffalo fell over the cliff, to the dismay of the expectant camp below. The opportunity had been utterly wasted.

And it was one of the very last opportunities. The future for the Blackfoot and all the Plains Indians lay not in buffalo jumps, but in adapting their techniques, traditions, and culture: following the herds as they crossed the plains, picking off individuals every day with arrows and, later, rifles—the "old way" of Sitting Bull, the new way of the horse.

So ended a tradition that had sustained the plains tribes ever since they arrived with their buffalo through the Bering Strait traffic lights thousands of years before. This long prehistory of jump hunting lies in evidence beneath the grassy scree of the jump itself. Archaeologists have dug down into the earth here and found that the thirty-foot drop that now exists used to be far longer—up to seventy feet. Broken rock, turf, and buffalo bones have accumulated here over the millennia, lessening the buffalo's fall. In between, and lodged within these bones, are Stone Age projectile points that chart the development in Plains Indian hunting technology over five thousand years. In the topmost margin, dating to about a thousand years ago, are Blackfoot arrow points, chipped stone heads notched on each side so that they could be strapped to the arrows. Below that, from eleven hundred to nineteen hundred years ago, are "Avonlea" points: thin, delicately crafted stone arrow tips that suggest a different tribal tradition. Among the bones below that, from nineteen hundred to three thousand years ago, are different points again: "Pelican Lake" dart points, from a time before the bow and arrow Idea arrived on the plains, when the weapon of choice was the atlatl, the spear thrower. And the lowest layer, fifty-seven feet below the lip of the jump, marking a time between three thousand and five thousand years ago, are "Mummy Cave" points: larger, notched dart flints. There are few in number, which suggests that at this time the tumbling buffalo were almost always killed on impact and the hunters didn't have to waste their weapons.

Were all of these cultural traditions the work of one people, the Blackfoot, or do their boundaries mark points in time when the valuable buffalo jump gained a new owner? It's impossible to say. Each of these point designs is identified with long-gone cultures in other parts of the region, but this doesn't mean to say that the Blackfoot and their direct ancestors didn't fashion these points here. As these goggles imply, Ideas live independent lives from people. What evolved elsewhere as a new Idea could easily have traveled to this part of the plains and settled into the minds of the ancient Blackfoot—that's how the Blackfoot whom Fidler met in 1792 had bows and arrows and rode horses. They hadn't invented the Ideas of the bow and arrow or the domesticated horse; they had adopted them, as all human minds can.

Standing at the top of the cliff, meadow birds chirping in the sun, looking down the jump to where a hundred thousand buffalo broke their necks, I finally realize that this freedom of movement between genes and memes means that I'll never be able to pinpoint which of the prehistoric tribes who came through this country first had the tepee Idea. It could have been the Blackfoot, or it could have been a completely different, long-gone plains tribe that dominated these lands before them. It all depends on how long the tepee Idea has been leaping between the minds of Native Americans. How am I supposed to discover that?

Humaneering

The final part of our story is the most exciting bit. Picture the scene: sometime in the last six million years, a new replicator, the meme, found itself at the wheel of a creature's evolution, able to exert control on every aspect of its voyage through design space, since meme-ability had become so securely associated with reproductive success. That said, the creature in question already had a lot going for it. It was a mammal, which meant that it had an expandable random-access brain. It was also an ape, which meant that it lived in social groups and had opposable thumbs, plenty of vocalizations, and an extraordinary power of imitation. It was a good model to work on. The memes couldn't afford to blow this opportunity. So, what would this "memetic drive," as Blackmore calls it, commission? What, in the best of all possible worlds, would memes "want" out of Life?

Simple, really. The same thing that genes "want" out of Life: to maximize their chances of survival and reproduction. From the point of view of memes, "survival" refers to the amount of time they are stored in a brain's memory, and "reproduction" refers to the events in which they seed near-identical offspring in another brain. So the creature that memes would "want" out of Life would:

1. Have as big a memory as possible, so that there's the maximum amount of storage space for the memes to survive in (design goal 1);
2. Be armed to the teeth with technology adept at passing on and receiving memes with a high degree of accuracy (design goal 2); and
3. Have an insatiable urge to use this technology to do so (design goal 3).

The creature that the memes ordered its genes to engineer has all three of these characteristics in spades. I'm one, you're one, Ads is one. Despite the fact that they began with an organism built for a totally different purpose, the memes have succeeded in their ambitions. So, how did they pull it off?

DESIGN GOAL 1: MAXIMIZE MEMORY

When you look at our closest living relatives, the species that haven't been driven by memes, and then you look at us, the species that has, one of the first things you notice is our huge, bulbous heads, giant in relation to our bodies. No other ape has such a disfigurement. These features are part of the solution to design goal 1: maximize memory. If you want a bigger memory, you need to build a bigger brain. How did the memes order such a big head? The neocortex was ready-built to enlarge, because it's essentially one tissue type, composed of the ever-expandable random-access circuitry. The only engineering issue came when expanding the skull. Bigger heads weren't the problem; giving *birth* to bigger heads was the problem.

Our babies have heads that are just under a quarter of their body length. That's about as big as a head can be when you consider that, at the end of pregnancy, it must exit a mother's body without undue risk to either the baby or the mother. On this count, the memes have pushed it. As I mentioned at the start of the book, childbirth is an extremely dangerous activity without modern medical assistance. Death during childbirth for mother and/or baby used to be commonplace. This is not something that the genes alone would wish for; it's no good producing such a high-cost fetus if it has only a medium-to-high chance of surviving the birth. It all smacks of memework to me.

Looking back through the fossils of our ancestors, I see that it could be that the memes have always forced this issue. The brain size of humans has taken a dramatic leap twice in our flip-book history—once about two million years ago, marking the arrival of *Homo erectus*, and again at the close of *Homo erectus*'s reign on Earth, five hundred thousand years ago, when our foreheads started to grow. It could be that these jumps were less to do with developments in brainware and more to do with adaptations to the female pelvis, enabling larger heads to be born with some degree of success.

As the brains got bigger, the memory got bigger, but it wasn't a simple quantitative gain. The expansion of the volume of the neocortex meant that the thoughts of our ancestors could become more and more abstract. The longer the neuronal pathways, the more distant the fireworks from the sense organs, the more processed the thoughts at the end of them. So for every centimeter cubed of added quantity, there was immeasurable added *quality*. The ballooning association areas were enabling our ancestors to build more context, concepts, and meaning into their lives. Size alone, without any special new hardware, would account for much of their growing intelligence—and, as a happy consequence, their disproportionately expanding intelligence created a disproportionately expanding range of niches for species of Idea. The enlarging skull space of humans literally enlarged the design space of meme Life.

There was another trick that the memes could pull. Simply by slave-driving a single gene that codes for the protein thrombospondin, every neuron in the neocortex could be stimulated to bump up its synapse number, increasing exponentially the number of connections between neurons. If memory and, indeed, memes are simply patterns of neuronal firings, then increasing the connectivity of neurons will radically increase both the memory and the meme count.

The memes spent six million years pumping up our brains, pausing to alter mum's pelvis at points, but then continuing with the program. The result was a creature that left the rest of gene Life standing. *Homo sapiens* had memory and intelligence like nothing else on Earth. We were in a world of our own.

DESIGN GOAL 2: DEVELOP MEME TECHNOLOGY

The capacity to remember Ideas is not enough. The memes also needed to maximize the proficiency with which their meme monkeys traded Ideas between brains.

The first job was to raise the general standard of imitation. Our recent ancestors were no monkeys, but even they were not that good at aping. Chimps, for example, may be able to copy a termite-stick, leaf-sponge, or hammer-stone trick, but complex psychomotor tasks such as tying shoe-laces are completely beyond them.

To improve the accuracy, or "fidelity," of meme transfer, there was a need to develop manual dexterity and precision. Luckily, an opportunity for this design work came along when our ancestors quit the four-legged approach to Life and stood up on their hind limbs. This left two front feet free to become hands, and the meme-gene coalition took its chance. In the intervening years, the anatomy of our front feet has changed so much that referring to them now as feet sounds, well, insulting.

But the memes didn't stop at clever hands. In their campaign to improve our imitative capabilities, they have enabled us to "read the minds" of others. This quality may simply be an extension of the mirror neuron capability that our primate ancestors developed—a deep mirroring of observed actions, perhaps—but what is certain is that when a human watches another human tying shoelaces, she is uniquely able to put herself *in that person's shoes*. She will be able to judge simultaneously and with some accuracy whether the person tying the shoelace is in a rush, anxious, distracted, or bored. This incredible social talent stems from what psychologists call a theory of mind, a capacity to understand that other individuals have their own minds, their own points of view, and their own desires and ambitions. This perception is so obvious to us that we expect all half-decent animals to share it; but experiments suggest that they don't. Not even chimps and dolphins unambiguously demonstrate "second-order intentionality," a knowledge of the intentions of others. And it's something that even we take a few years to figure out.

If you show a three-year-old a tube of Pringles and ask her to guess what is inside it, she will probably say, "Pringles" (if she is suitably indoctrinated in the leading brands of snack foods). If you then show her that there are in fact only pencils inside the tube, she will be surprised but will make the mental adjustment. Then, if someone named Steve enters the room and you ask your three-year-old, "What does Steve think is inside the tube," she will say, with all confidence, "Pencils." She doesn't appreciate that Steve has a different experience of life than she. She doesn't understand that Steve has a mind of his own. Do the same experiment when she's four, and she'll probably get it right; her theory of mind will probably be in place by then.

The genetic advantage of "mind reading" in this way is clear: it allows individuals to judge the trustworthiness of others, which is an extremely useful talent for any organism reliant upon the cooperation of others to survive. But mind reading is, of course, highly beneficial for memes, too. Understanding another person's intentions while they are performing an act will certainly help an imitator avoid copying *unintentional* acts, and any extra insight into the working of another's mind can only benefit the fidelity of meme transfer.

The dawn of our "theory of mind" might have occurred at any stage in the last six million years; our ancestors' burgeoning capacity to mind-read didn't leave any fossil evidence, so we have no way of knowing. But some psychologists maintain that a theory of mind had to be in place before the evolution of the next great meme technology: symbolic communication.

A Symbol Creature

A revolution in meme Life, something akin to the accidental onset of multicellularity in gene Life, ensued with the advent of symbols. A symbol is something that, arbitrarily and by convention, is associated with something else. A thumbs-up sign, for example, is a particular type of symbol: a gesture. It's completely arbitrary; it doesn't have any value in life independent of its meaning, and it could easily be replaced by a different shape of the hand if only we all agreed to go through with the change. But as an accident of our species' social history, the thumbs-up sign has accrued a conventional meaning. In fact it has many conventional meanings, depending upon the circumstance. In everyday life, a person giving you a thumbs-up is communicating "good?" or "good," or "okay?" or "okay." However, if that same person is underwater and wearing SCUBA equipment, the gesture stands for "I'm going to the surface." And if he is standing by the side of a road with a backpack, it means "can anyone give me a lift?" And if he is hitchhiking in Iran, whether aware of local convention or not, he means "up yours."

This facility for synchronized Idea association offers carte blanche to meme Life. Any number of more and more abstract Ideas can become anchored to a symbol as long as everyone agrees to take away the same (functionally identical) meaning. So it's perhaps not surprising that our meme-driven ancestors would ultimately seek out a system of symbolic communication that maximized opportunities for Ideas, a system based upon a different type of symbol, a sound symbol: the word.

In essence, words are similar to gestures. A gesture is an imitable sequence of muscle contractions in the face or hand, with its associated meanings. A word is an imitable sequence of muscle contractions in the tongue, lips, mouth, and larynx, with its associated meanings. Both are arbitrary, and both require a conventional association to be of any value. So they use the brain in the same way.* Indeed, recent work[2] has discovered that symbolic gestures and spoken words are coordinated by

---

*Hence words are simply another form of Idea: with each individual word and, indeed, each part-word, subject to natural selection and cultural evolution just like any other Idea. This explains why McWhorter cannot recognize Darwinian evolution in languages (see chapter 11). Languages are a level of organization *above* the cultural species—massive communities of living words able to lose and gain new species of words freely—which is why they appear to split and merge like clouds.

the same parts of the brain—the famous Broca's and Wernicke's areas on the left side of the cortex—growths that have long been recognized as unique to humans, and hence hailed as the key to our uniqueness. But, ultimately, it was words that would come to dominate our symbolic communication, and take it to a new memetic dimension. This is because words have various important advantages over gestures. You can use words when your hands are busy. You can speak words to someone you can't even see, or who can't be bothered to watch you. The fidelity of words is greater than the fidelity of 3-D hand movements, so their meaning can be received with greater accuracy. Employing the tongue, lips, and mouth cavity in different ways, you can manufacture numerous different sounds to bring variety and, more important, *expression* to words. Yet delivering these many different words takes up only a little of your energy, and a little of your time; you can spew out words, one after another, at quite a rate. So, when the first language arrived on the scene, while it could have been composed of other symbols—gestures, expressions, whistles, or even dots and dashes—it's perhaps understandable that it was words that won the contract.

A language is simply a system of using symbols in such a way that the position of each symbol in a sequence has a bearing on the meaning of the other symbols in that sequence. The trick is in developing a convention of symbol-ordering rules, or grammar, in the first place, and there remains a significant debate among researchers about how we each come to do this. How do we collectively know that the sentence "man eats dog" and "dog eats man" mean different things, even though they contain the same symbols? For much of the last fifty years, psychologists and linguists have argued that this talent is so special and so complex that it must be under the jurisdiction of an innate capacity, a language acquisition device. However, recently this view has been challenged. Perhaps, say the critics, language is just a feat of superb memory. Maybe we simply piece together the conventions of grammar as we grow up listening to a language of sound symbols and consign it all to our massive memories. After all, we're good at learning conventions.

If so, then all those words and all those rules of grammar are no different from everything else we "know." They are simply inheritable Ideas of one form or another, logged away in our amazing memories, linked, carefully and over many years, with lots of others according to our mutually agreed conventions. Our entire cognition becomes no more than a vast and magnificent house of cards, each knowable "thing" gaining its value, its meaning, solely as a consequence of the links it has to other knowable things. Our intelligence becomes a spectacular artifice of neuronal firing patterns—of firework displays—each one wired up to others and eager to trigger them. And since, at least through these goggles, each of those firing patterns lighting up the cortex is a meme, the entire content of our minds becomes a melee of memes inhabiting the space as an energetic, writhing, fundamentally interdependent world of Ideas.

We can each grow this amazing noospherical ecosystem, a jungle of wild Ideas in our heads, only as a consequence of our expertise at symbolic communication. When you boil it all down, our capacity to receive or pass on those deep Ideas relies entirely upon our ability to move our tongues or fingers at the right time and in the right way. Those mutually-agreed-upon symbols hold the key to the vast proportion of meme Life, of our cognition, to such an extent that when a child is found who has been denied the process of "enculturation" during the first few years of her life, authorities find it near-impossible to rewire her already-mature brain to suit our world. She, a "feral child," has all the right adaptive features, all those characteristics that make her the ultimate meme machine, but without a grounding from a very young age in a system of symbols, she will never be able to host complex Ideas.

Reason to Believe

DESIGN GOAL 3: MAKE THE MONKEY TALK

So here's the status quo: a fancy monkey, its brain infested with a new Life Force, goes through millions of years of radical physical and social change, commissioned by its misguided genes, in order that the new Life Force, meme Life, can have a larger stage to play on and a larger audience to play to. But how do the memes get the fancy monkey to put on a show?

We each have a vast memory space buzzing with Ideas, and the most incredible space-age technology on standby to enable us to pass those Ideas on to others (no matter how abstract they are). But if we have no urge to use this technology, if we'd rather spend our time eating fruit, picking parasites out of our neighbor's hair, and having sex, then, from the meme's-eye view, it's all for nothing. Of course natural selection wouldn't stand for that; the memes wouldn't stand for that. They've made certain that we are insatiable consumers, producers, and disseminators of culture, unable to stop ourselves from searching out new Ideas, chucking together novel meme combinations in our spare time, and talking about them nonstop. We can't help it. We gossip, we tell stories, we teach, we ape; we're inquisitive, we're passionate about a cause, we can't stop thinking about things, we can't keep a secret; we understand each other, we believe in ourselves, we are cocky and confident. Just as plants are addicted to the sun, and animals to plants, we are born Idea junkies. If we find ourselves alone on a desert island, unable to get an Idea hit from someone else (. . . anyone else), we quickly descend into cold-turkey madness. We mumble to ourselves. We twitch in our sleep. We lose the will to live. We obsessively tie and retie imaginary shoelaces.

I'm not intimidated by this reality. I don't mind being addicted to Ideas. In fact, I quite like it. They've made me like it. I enjoy absorbing them, rekindling them when they're dormant. I'm happy to pass them on, be it special Crow ties or ordinary shoelace ties. In fact, it's more than that. I *need* to pass them on. That's why I'm writing a book.

So does that mean I'm an egomaniac? Is it a psychological abnormality that makes me believe that others out there want to hear or watch or read about my Ideas? Well, you could say that—it is abnormal in the biological world—but among the members of our species, in our world, egomania comes as standard. We all have incredible self-belief.

"And why do you think that is?" say the more teasing of the goggle makers, but it's a rhetorical question: "because the memes want you to believe in your Ideas. If you believe in them, you're more likely to feel the need to relay them to others." You have to agree with the logic. Having a self, an "I" whom you identify with, wrapped in its own personal convictions, does make each of us more likely to spout our Ideas, exactly because we then think of them as *our* Ideas, part of our unique makeup, something to defend. But, say the goggle makers, the "I" in you, is an illusion, part of the grander memetic artifice. Could we be deceived to that degree?

There's some reason to believe so. A series of experiments, terrifying to some, seem to suggest that we have no objective free will. Researchers have found that when volunteers are asked to choose to press a button with their right or left hand, brain scans can predict their decision up to ten seconds before the volunteers have the impression that they have made the decision.[3] It seems from this experiment that our conscious selves are not so much in charge of proceedings as much as they are a *record* of proceedings, a memoir of the predetermined actions of our subconscious.

This is shocking news. It pulls apart the dearly held belief that we have agency in our lives. The safest thing is simply not to believe it—but then, you haven't heard about the fallibilities of your belief system yet . . .

A new generation of cognitive psychologists less enamored with our psyches than their forefathers is beginning to discover just how poor we are at choosing exactly what to believe. To begin with, we are, in the main, incredibly gullible, because our higher cognition is essentially a botched extension of our perceptual systems, and we rarely have to question whether those systems are telling the truth. We have a distinct tendency to believe the first thing we're told over any subsequent version, because the first version of the truth will always be the one that we most rehearse in our brains. We are suckers for a good sell. If we find the proponent of an Idea charismatic, then we are far more inclined to believe in it than if we don't take a shine to that person. Likewise, the mood we're in when we receive a new Idea has a strong yet irrational influence on our acceptance of a new truth. Despite the partiality in play, we then rarely remember accurately the source of the "truths" we take as gospel—because our semantic (meaning) and episodic (personal event) memories are separated in our brains—and yet we don't appear to care that our referencing system is screwy. We are riddled with a "confirmation bias," a strong propensity to "believe" only those Ideas that support or at least loosely collaborate with our existing beliefs. To add to that, we indulge in energetic "motivated reasoning," picking apart others' beliefs if they conflict with our own.

Don't believe me? Just look around the world. There are people who believe all sorts of things—that there are ghosts, that the Earth is flat, that black cats are unlucky, that there was no Holocaust, that their particular way of tying a shoelace is *the best*—ludicrous, silly things, simply because they were the things they heard first or that best fit another silly thing they've heard. And this faulty belief-acquisition system leaves us all compromised. As we fill up with a particular flavor of Idea, we all become fundamentalists to some degree, psychologically incapable of taking an objective view. Stubbornly convinced that there is a God or that there isn't a God. Or that there is a self. Or that there isn't. I don't know. I don't know what to believe anymore.

But I can see a reason to believe; the reason is that the memes wish us to. With enough zany cultural foundations, faulty belief systems, and adamant selves, there are willing, convicted, chatty advocate hosts out there for practically any Idea. Somewhere in the human world, every Idea can find a home.

Among Life on Earth we *are* weirdoes, but we're not *inexplicable* weirdoes. We are a species that has been crafted by two replicators. Our first two thirds were built by cranes commissioned by the gene. The last one third was built by cranes commissioned by a replicator unique to our species, the meme. The mystery of our past is solved: no skyhooks were used in the making of this species.

The Ghost of an Idea

The wind picks up on the plain below Head-Smashed-In. It's late in the afternoon and I'm marching out alone on an exploratory trail with a free leaflet, excited because I was told by Kathy, one of the Blackfoot guides in the visitor center, that I may just find what I'm looking for out here among the grasses. I get to the far end of the trail, a good distance from the kill site. The breeze fills my ears with white noise, and I feel alone on the plains.

It must be around here somewhere. I pace back and forth looking for clues, impatient; I talk to myself; I kick at the swards. Then I get down on my knees, out of the wind, and feel among the grass blades. They are thickly packed, warm, sharp, and stiff. Grasshoppers spring off in all directions like mortars from a silo. I crouch low and sweep my arms left and right at full stretch, in a vain effort to survey more ground. I feel nothing. Then I move to the left and scythe the tussocks again. I repeat the move over and over, getting more and more frustrated, more and more speedy, more and more slapdash. My chin brushes the grass tips. I breathe in grass dust, contorted in an awkward stretch. The wings of insects tickle my cheekbones, and the skin on my arms begins to itch. Then, as tears start to flush the pollen from my eyes, I touch something with the fingertips of my left hand: the cold hardness of a stone.

I leap over to it, on my knees, and part the grass around it. It's a large, smooth, gunmetal gray stone buried in the grass and partly in the soil below. I crouch-walk off to the left of it. There's another. And to the left of that, another, and another, spanning the ground in a tight arc. I stand up and use my foot to sweep onward, pushing the grass down as I go: bang, another . . . bang, another . . . Soon, I've traced the entire circle.

"They were like reserved parking spaces," Kathy had said. "You wouldn't dare pitch on someone else's." What I've found is a tepee ring, all that remains of a campsite that used to exist here below the jump. These stones once held down a tepee cover. They were cast aside one moving day, when the camp was collapsed in half an hour and Indian women pulled the tepee cover off its poles. The stones collectively rolled outward as the cover was withdrawn, leaving a ring slightly larger in diameter than the tepee itself. As the snow melted in the spring, the stones sank into the mud beneath, deeper and deeper over the years, becoming part of the landscape. They are the trace of a tepee, the only evidence that it ever existed.

An archaeological survey of the high plains in the 1930s[4] unearthed 858 of these sites, old camps used year after year by the tepee dwellers. The best preserved have more than a hundred tepee rings. Some rings overlap. Some have sunk deeper into the prairie soil than others, revealing their age. The camps were often arranged as large rings themselves, each tepee set with its doorway facing east, toward the morning sun (and away from the prevailing winds). The large camp circle in turn had its own "doorway" facing east: a gap in the lodges through which people would enter and leave the camp.

The rings are typically between twelve and eighteen feet in diameter, small compared to modern-day tepees. Their size and the fact that they were secured with stones suggest that they were chiefly buffalo hide tepees from the dog days. Buffalo hide tepees were much heavier than the later canvas ones, so they tended to be smaller, especially when they were being dragged about the plains by dogs. In addition, canvas tepees were rarely held down by stones, because if the canvas came in contact with the wet ground, it would rot; they were usually pegged to the earth with laces.

But attempting to date these sites, beyond an association with the dog days, is notoriously difficult. Albertan archaeologists prefer to spend their time establishing a chronology for the knapped tools they find nearby. The buffalo jump, as I explained, was found to hold weapon points dating back five thousand years, but down here, below the jump, digs have uncovered much older relics. "Scottsbluff points" are large worked stones that are stemmed at their base, meaning they were hafted onto spear shafts. The "Cody culture" that crafted these points did so nine thousand years ago, in a time before even the spear-thrower was invented. With only heavy spears at their disposal, you'd imagine that the Cody were even more keen to use the jump. But there is no sign that they did. Perhaps this was because they were living during the Younger Dryas, a particularly cold period, when the glaciers started to creep south again and the buffalo herds may have been scarce in the northernmost plains. Perhaps it's because the buffalo that did graze here would not brave the chilling winds on the high ground above the jump. Or, maybe, and most likely, it was because the plains people just hadn't yet thought up the buffalo-jump Idea. Maybe it was only upon the invention of the spear-thrower—a device that could pierce a hide from some distance and spook a herd enough to muster a blind stampede—that the first buffalo fell off a cliff (accidentally) and someone watching thought, "A-ha!"

The tepee ring I'm sitting in probably doesn't date as far back as nine thousand years; at sites where archaeologists have been able to put a date on the rings, they are found to be rarely more than three thousand years old. But that doesn't necessarily mean that there were no tepees before that, and I'll tell you why.

Far to the northwest of where I am sitting, back up the corridor between the mountains, across the high country of Alaska and over the currently sunken occasional land of Beringia, tribes of Palaearctic people live on the northern tundra and arctic coast of Siberia. These days they tend to reside in prefabricated buildings, and work in the oil or mining industries or herd reindeer, but not too many generations ago, before the Asian buffalo herds suffered the same fate as their North American cousins, these tribes hunted buffalo. In the summer, when they ventured out onto the plains to hunt, they took with them temporary dwellings, variously called *chum, yaranga*, or, way over in Scandinavia, *lavvu*. If you visit these people and ask nicely, they might fetch you one to look at; most villages keep a few for small hunting trips or traditional celebrations. What do they look like? They are conical tents made from hide or canvas and supported by a skeleton of poles, with a space for fire smoke to escape at their apex. They look very much like tepees.

Those are the tents that would have come over with the ancestors of all Native Americans when they broached the traffic lights of Beringia all those millennia ago. It was those tents that were dragged down the corridor of ice and onto the unpeopled plains of North America. It was in those conical tents that the first people of America were born.

They were not tepees, exactly. Existing *chum, yaranga*, and *lavvu* don't possess those two critical tepee characteristics: smoke flaps and an asymmetrical cross section. But they sure look like the ancestors of tepees, an Idea from a relatively recent page of the tepee's ancestry flip-book still happily alive in thousands of minds from one side of northern Asia to the other.

So, between the arrival of the first people into America and the arrival of the first Europeans on the plains, the tepee with its two critical features came into being. Though all the tents in between were presumably also weighed down from the wind and rain with a ring of stones, we find very few old tepee rings, making it hard to pinpoint the date of their arrival. Perhaps tepee rings just take about three thousand spring snow melts to vanish from sight.

My guess is that the origin of all tepees occurred a few thousand years before the oldest tepee ring, in the centuries following the initial buffalo-jump Idea. That would have been the first period in which buffalo hides were plentiful. It would have been the first time that the local people could afford to swap their hide tents for new ones every three to five years; the first time they could spend their entire lives up here on the open plains and remain well fed and clothed. Armed with their new spear throwers and their new buffalo-jump Idea, they may have been the first people to live year round on the high plains eating buffalo. A permanent life under the hide would have forced improvements in the design of their tents. Formal smoke flaps were a possible first step; many of the Asian tribes throw a blanket over the hole at the top of their tents to control smoke flow, so an incorporated hide flap is not much of a leap. Next was the drive to asymmetry, which we were still seeing in relatively recent times as the three-pole replaced the four-pole structure.

It's bound to have been a gradual tepee dawning—shuffle steps the whole way. That's the way evolution is. So gradual, in fact, that it's impossible to spy an origin at all. Every conical tent Idea in the tepee's ancestry flip-book would have been able to "breed" with the one that came before it and the one that came after it. It's only after the event, in retrospect, that you can declare, "Hey, it's happened." Because Ideas *are* like species, there was no definitive "origin" of tepees.

I catch the slight sound of a whistle and glance around to see what must be Ads in the distance, waving. The figure points at his watch, a symbol that I can't mistake. Quite right. It's time to leave the past behind. I'm at peace with it. On to the present.

15

The Present

Welcome to the Jungle

In the pleasing half-light of a midsummer evening we find that we have returned to proper freeways, where there is a proper volume of cars. There are expensive city cars, BMWs and Mercedes, cars that have no place on the Great Plains, where an automobile's life is counted in dog years. There is an increasing number of humans living by the side of the freeways. We're approaching Calgary. After two weeks voyaging through a remote cultural archipelago, sailing among a string of sparsely populated islands, this feels like a return to the mainland. Ahead of us are one million minds, all in one place. One million minds engaged with the very latest version of meme Life. One million minds collectively hosting the present. That's all "the present" is, and if we were all to forget it tomorrow, then—poof—it would be gone.

"Right, RIGHT!" shouts Ads. I spin the wheel. We are fed down a slip road and onto a straight street with intersections going off to the horizon. Overly garish colors, oversize logos, overly bright neon lights fill the sky at all heights—there are chickens and cowboy hats and cartoon characters and maple leaves, each jumping up and down, shouting, "Ooh, ooh," desperate for our attention. My symbolic mind goes into overdrive. An unruly hoard of memes begins to tickle my consciousness. I'm being asked to recall associations I have with fast-food chains. I'm being asked to remember what a "lube shop" is. I'm being asked to work out where exactly "drive-in wedding chapel" should fit in my Idea community.

The sunlight finally vanishes. The traffic gets thicker. The buildings get taller. By Seventeenth Avenue we've entered some kind of evening promenade on wheels. We creep along in the Chrysler. Horns are blasting, lights flashing. Motorbikes fly by, weaving in between the traffic. One guy in a German army helmet on a hog bike takes pleasure in roaring the accelerator past the clientele of the blinking, thumping bars that line both sides of the road. It's Friday night, Calgary's youth are in a good mood, and all we can do is roll down the windows and enjoy the parade.

There are sneakers hanging from telephone wires. There are jeans riding low on waists. There are different pant cuff heights, different ways to say hello, different brand names on shirts, different colored dyes for your hair. Young guys debate hockey in a language of numbers. Girls discuss their favorite soap operas. Bar staff mix cocktails. People gossip over bar tables, between cars, across the street at high volume, on cell phones. Everywhere I look, my goggles are dazzled by an astonishing profusion of meme life.

I can imagine how it continues. In hotels, maids will be folding triangles on the ends of toilet rolls. On verandas, grandmas will be using knitting patterns handed down from their grandmas. In community centers, there'll be talks on a trip to Machu Picchu or the suffragist movement. Teenage girls will be reading their vampire thrillers. Children at sleepovers will be telling one another ghost stories in the dark. Tina will be cramming for her science exam. Bob will be checking Google for alternator problems. Sasha will be learning "Clair de Lune." Paolo will be teaching some Italian to his friend Pete. Gabriel will be dreaming about the few lines he has to say in the school play. And Eddy, over the dinner table, will be telling that joke again about the two television antennas that got married.

How do I make sense of this mesmerizing memetic world? How can I possibly hope to catalogue the meme Life alive here within this million-strong meme-monkey jungle?

Relax, I'm a trained ecologist. Making sense of jungles is what I do best.

In fact, let me tell you about jungles—real jungles, not cultural ones. Jungles, or, more properly, tropical rain forests, are the peak of gene Life. They are the most productive, complex, dynamic, and diverse ecosystems on the planet. The first Europeans to witness them were numbed by their irreducibility. "We rush around like the demented," said the German naturalist Alexander von Humboldt in 1799, "in the first three days we were unable to classify anything; we pick up one object to throw it away for the next. Bonpland keeps telling me he will go mad if the wonders do not cease."[1] But the naturalists kept their heads and worked tirelessly on deciphering the apparent chaos. Two hundred years later, tamed by the science of ecology, the rain forest is finally beginning to make sense. This rationalization doesn't take anything away from our wonder; if anything, the jungle is more wonderful than ever.

We now understand it as a three-dimensional stage that, day and night, 365 days of the year, hosts the most convoluted theatrical performance in the natural world. It's a production with countless story lines, numerous relationships, and elaborate character histories. There's a cast of millions, with some well-known stars, but the vast majority of the players are completely anonymous, carrying out bit parts with subplots opaque to us.

The basic plot is simple. Each of the many millions of animal, plant, and microbe species must secure unique solutions to four universal problems: they each need something to eat, somewhere to live, some way of protecting themselves, and somebody to "love." The rain forest is unique among ecosystems in that the opportunities for solutions to these problems are almost endless. This is partly because the architecture of the habitat is so complex and the species diversity so high, but it's also because the sheer density of living things, with species living upon, in, and on other species, creates a playground for natural selection. Evolution has crafted the most astounding relationships here, associations that over time draw different species closer and closer together. "Coevolution" (when two species adapt in tandem) and "specificity" (relationships with profound and particular commitments) are the norm. And because most species are involved in more than one of kind of relationship, the whole forest has unwittingly signed itself up to a comprehensive treaty of inter-dependence.

Take one solution for each of the four problems:

▲ SOMETHING TO EAT. Forest elephants in the Congo go out of their way to find the fruit of the omphalocarpum tree. They know where every fruiting tree is, and they build trails as wide as bridleways through the densest forest to get to them. The omphalocarpum fruit is a formidable meal. It's the size of a bowling ball and about as hard. When it ends its fall from the canopy it makes a low-pitched thud that travels well through the forest, announcing dinner. Elephants come from miles around to feast, and after the meal the elephants head back down their trails, depositing the omphalocarpum fruit seeds in huge "grow bags" of dung as they go. Generations of elephant traffic have lined the trails with omphalocarpum trees and the thirty other fruit trees that rely exclusively on elephants for their dispersal. So these trails are no longer just routes to fruits, but open marketplaces with a bounty of goods on offer.

▲ SOMEWHERE TO LIVE. In the breeding season in Borneo, pairs of great hornbills take up residence in the holes of old canopy trees. Each site has to be selected carefully, because it becomes the female hornbill's prison for several months. In order to protect the chick from predators, she barricades herself in the hole with a mixture of clay and regurgitated fruit. The male hornbill is then charged with visiting the hole several times each day to feed the female and, after its hatching, the chick. The female leaves the nest when the chick gets too big, at which point the chick reseals the nest and remains inside for another month before fledging. Property is in such demand in the canopy that the number of old, holey trees determines the number of hornbills in an area.

▲ SOME WAY OF PROTECTING THEMSELVES. Ants are vicious things; they bite anything that gets in their way. But their foul temper has been put to good use in the rain forest. Many plants enlist ant colonies as security guards, housing them in lavish hollows within their stems and feeding them with sweeteners from special nectaries. In return, the ants defend the plants against leaf-eating insects. But it appears that the system is open to corruption. Some caterpillars are equipped with organs that deliver a "backhander" of proteins to any ants that come near. The ants accept the bribe openly and not only let the caterpillars get on with their leaf munching but also keep a lookout for the predatory wasps that plague them.

⚘ SOMEBODY TO "LOVE." Orchid bees look like jewels as they fly through the Amazon: their bodies are covered in a dazzling metallic green or blue. But this livery is not for wooing others. Male orchid bees use pheromones to attract a mate. They make these pheromones out of a range of rare compounds offered up by orchids that grow in the canopy. But it's not a free gift: the orchids hold the bees for ransom by releasing only small amounts of these chemicals at any one time. So the male bees have to be patient, repeatedly visiting all the orchids in order to accrue enough compounds to go a-courting. In toying with the bees' love lives in this way, the orchids manage to carry out their own romances.

Intimate, specific, but consequential relationships exist in the rain forest in such proliferation that we are still, two hundred years on, only just scratching the surface. There are so many ways to make a living, so many niches available, that in truth we will never learn them all. Yet the consequence of this tangled web is easy to understand: No living thing in the rain forest lives alone. Every organism is part of a remarkable community of Life, struggling to find a little space in which to make a living.

In all of these respects the city of Calgary is exactly like a noospherical rain forest, a cultural jungle. It is a mad chaos of Ideas living upon, in, and on one another. Every Idea must secure its niche. They are all in competition with one another to find their way into our minds, and to some extent our hearts, to become important to us. Over time, in pursuit of these ideas, they will adapt to best suit their particular role, and more often than not, nurture intimate, specific, and consequential relationships with other Ideas. The ways in which they achieve their goals will be wide and varied. If I attempted to catalogue them—to classify the cultural equivalents of orchid bees or omphalocarpum—I'd end up like von

Humboldt and Bonpland: rushing around like the demented, picking up one item only to throw it away for the next. I haven't got the time, space, or energy to make a species list, but I do have a few ideas of how to make some sense of all this wonder.

Idea Ecology

Ads and I find a corner seat in the Lido Café, "an institution," it says in my guide book. It's a greasy spoon run by Chinese owners with fifties booths, vinyl seats, Formica tables, and table jukeboxes, but the lack of effort to polish the look with 100 percent retro décor makes it feel rather authentic. We had a few beers after checking in last night, so Cokes, coffees, and a big fried breakfast are in order. I get sausage, scrambled eggs, and hash browns. All three are different species from those endemic to England: the sausage is a disc of fried meat, the scrambled eggs come as a smooth yellow hubcap, and the hash browns are sautéed potatoes. No matter, they taste great, as long as you add ketchup. I put my ketchup on the side; Ads is an all-over-the-plate kind of guy. The differences in each of us only serve to give those little Ideas even more room to diversify.

Next to our table is a young French Canadian family. Mum and Dad part-engage in an ongoing conversation, while Mum repeatedly picks rolling crayons off the floor and passes them back to her son (who's coloring in a picture of two dinosaurs), and Dad attempts to feed brown juicy apple pulp to their baby daughter. Baby's having none of it. Despite being restrained by a café-standard high chair, she's able to move like a street dancer, twisting and turning, flaying her arms about, and turning her head away, lower lip folded over the top one, eyes tight shut. Progress is slow.

Then something changes. Dad draws on experience and hands her the menu. She stops, stares at it, then feels its edges. He moves in closer with the yellow spoon. She unfolds the menu and gazes at the squiggly lines. As the spoon gets close to her pursed lips, Dad can't help but open his own mouth wide. She catches the movement, stares at his lips, and can't help but do the same. She *has* to imitate him; it's involuntary. The fruit gloop goes in, and although half of it comes straight back out again, the delivery is deemed a success. Dad gushes verbal rewards in his best baby French.

I can't help but stare. The similarities between the gestures and expressions of these French Canadians and the gestures and expressions of French *French* are startling. I've spent a lot of time in France, and this family could be French. It isn't the language; it's the body language. They raise their eyebrows, pout their top lips, move their hands, and even shrug—they actually shrug—just like the French. This entire approach to communication is absent three hours south of here. Those psychomotor Ideas, subliminal movements and expressions, and of course the French language have not made it south of the border. Ideas like these need to surround you during your childhood to be adopted. Inside the mind of the boy sitting at that table—subconsciously listening to the words of his parents, glancing up to clock another unnoticed observation of their hand gestures and facial expressions—familiar fireworks displays will be flashing among the mirror neurons, excitations not powerful enough to create a muttering or action in him, but powerful enough to reinforce the synaptic connections he needs, to prioritize those patterns for when *his turn* comes to do so. That's what childhood is: a period of *culture practice*, a phase in which Ideas can jump on board even as the brain matures. Human beings have the longest childhoods of any animal. Our bodies, including our brains, don't finish growing until we're into our late teens, even early twenties. And, in a cultural sense, the period of childhood is still growing—rarely do our sons and daughters drift off and have their own children when they get to sexual maturity. In more developed countries it is now normal for children to remain at home even when in their twenties, perhaps because there is so much culture to inherit before they leave.

One way to make sense of the cultural jungle is to chart what happens during this extended period of enculturation, when it creeps in to germinate and flower within a single growing mind. That baby girl stubbornly refusing to eat has a near-blank mind—almost memeless. But it won't be so for long. Over a period of twenty years, she will inherit a mass of memes from the French Canadian jungle that surrounds her day and night. And she'll do so in a particular and predictable way.

Back in university, studying ecology, I learned how, given enough time, a patch of bare rock would transform into a bountiful jungle. It happens through a process called primary succession. The first step is for the lifeless rock to be colonized by primitive "pioneer" species—cyanobacteria, lichens, and free-living algae—which break down the very surface of the rock by oozing chemicals directly onto it, and begin to build a primitive sliver-thin soil. Microorganisms—bacteria, amoebae, and others—migrate onto this living film and set up shop, consuming the photosynthesizers and one another. As time goes on, the species diversity rises, the community becomes more complex, and this gaining complexity opens the door for yet more species. Minute worms, mites, and insects arrive. Food chains are set up. Fungi infest the substrate to feed on the mounting carcasses.

The living community grows in number, volume, and stature. A soil of dead bodies and cast-aside leaves gathers on top of the rock. This deepening substrate enables larger plants to take root, which in turn allows larger animals to make the community their home. And this is how it will continue. All but the pioneer species require other species to go before them to lay the ground. They cannot survive without the preexistence of the species they live in, or prey on, or work with, or parasitize. So it is a positive feedback situation: the more the community grows the more it can then grow.

The thin green layer becomes a sward. The sward becomes a thicket. The thicket becomes a bush; the bush becomes a woodland; and the woodland becomes a forest. This stockpiling of gene Life is a continuous process, but ecologists cannot resist the temptation to categorize the passing communities into "seres." They talk of early seres (the thicket) and late ones (the woodland). However, ultimately the shuffle in community members will cease. The soil becomes as deep as it will ever get, the plant life as glorious, the animal life as abundant as it can be. The succession ends in a final "climax community." Species may still arrive and replace those present, but these are like-for-like swaps; the climax community itself is stable and complete.

This is how even the most byzantine jungles are built. And, I suspect, it's how even the most byzantine minds are built, too. The baby French Canadian next to me is, even now, garnering her pioneer community, a foundation for all subsequent Ideas. At first they will be the simplest of simple psychomotor Ideas, aped physical behaviors. She'll copy the way her dad picks up a spoon. She'll copy the clapping hands and the smile and the laugh. She'll copy a forced blink, an open mouth, and peek-a-boo hands. She'll copy all of these things because human beings are preprogrammed to copy just about anything they can.*

---

*That's not to suggest that the mind of this baby French Canadian, or indeed any of us, was once a blank slate, or tabula rasa, that her brain was born featureless and inconsequential. Just

Her first attempts at language will also be psychomotor—"da-da-da-da," "ya-ya-ya-ya"—inherited phonemes and syllables. They are merely copied mouth and throat movements, but put together, and in combinations, they will someday deliver for her an entire language of words.

The important stuff behind the words, the symbolic associations, are a different class of Idea. I call them semantic Ideas, inheritable meanings. They are not directly tied to muscular movements. They exist solely in the brain as abstractions: specific fireworks displays in the neocortex that can be linked to the psychomotor Ideas via association areas. They are later successional species: they can arrive only once the right pioneer psychomotor species are in place.

The semantic ideas are a vast kingdom of meme Life with seemingly endless variety and complexity. They are super^n-organisms living in tight symbiotic relationships with one another—so tight, in fact, that it's often difficult to distinguish them. But the more a mind's living community of Ideas grows in number, volume, and stature, the more semantics it can host. The young boy coloring in the dinosaurs has clearly already inherited a string of more and more complex semantic Ideas. He knows that the subjects of his pictures are, first, animals (evidenced by the way he moves the page from side to side, pretending they are stepping), second, dinosaurs (evidenced by his roars to animate their drama), and third, one carnivore and the other a herbivore (evidenced by his choice of colors: one in green, the other blue with red drops falling from its jaws). He's inherited all these super-Ideas perhaps as a result of his father's tuition, perhaps due to his own concentrations and deductions, perhaps subconsciously. And as new, ever-more-advanced Ideas arrive to join his mind community, his comprehension of this meme-built world becomes ever richer. For example, he knows, now that he is four, that the two little lines suspended over the knees and tail of the Iguanodon are not flying worms but marks that symbolize the movement of the terrified prey (he goes over them in red while roaring). The boy's growing community of semantic Ideas works in cahoots with that uniquely fashioned human brain, such that, increasingly, he's able to read the codes of meme Life.

---

as any patch of rock is imbued with characteristics that will dictate the nature and direction of its primary succession, every human brain has its own unique "geology" that exercises influence on the coming enculturation. And thank goodness for it.

At four, he's still got a long way to go. To operate memetically within the modern world and to play a proper role in the noosphere, he will need to commit at least fifteen more years to the Idea succession. Society has formalized the majority of this enculturation within a system of compulsory education. The moment he enters the classroom, trained professionals will ensure that he starts his journey by encouraging him to accommodate a large assemblage of Ideas associated with the *writing* of language. Again, as with spoken language, the foundation of writing is a suite of psychomotor Ideas: pen strokes. Again, these are tied to other Ideas: pronunciation and punctuation codes. And ultimately, this odd and involved manual craft is connected to the same word-association fireworks as the spoken form of the language. And it is connected by way of the same part of the brain: Wernicke's area.

Exploiting the boy's newfound literacy, formal education will then undertake the task of building up his Idea community one species at a time, with due regard to the rules of succession, ensuring that early-succession species (e.g., the diameter-of-a-circle Idea) arrive before late-succession species (e.g., the πr^2 Idea). En route, the boy will become host to a string of new psychomotor Ideas (e.g., the way to hold a tennis racket, the mouth and throat movements required to pronounce Spanish words, how to use a mouse), and a whole universe of new semantic Ideas in math, history, music, science, and all the other fields. His capacity to efficiently house this long line of colonists—to host the growing cultural rainforest—is reflected in the measure we call intelligence. If he is particularly good at hosting semantic Ideas, he'll probably have an appetite to stick at education beyond the required age, to continue formally accruing his Idea community well into adulthood.

Why do we go to all this trouble? Why should we endure all these years of schooling? Through these goggles, the answer's obvious: because it suits the memes. Take, for instance, the enforced craft of writing. Writing is a wholly cultural artifact—it's made entirely of memes—and it's hard work to learn because it was invented so recently that our brains have never had the opportunity to evolve any hardware or software to deal with it. But from the meme's-eye view, literacy is a must, worth all our efforts, and that's because it offers Ideas eternal life. Any Idea that manages to get written down can potentially live forever, dormant on a page, in a document on a hard drive, or, indeed, laid down in stone, until the day another human mind reads it, whereupon it springs back to life.* So a population of literate human beings is the Holy Grail as far as memes are concerned. Just as it is in the meme's benefit that we force-feed our young Ideas for fifteen-plus years in classrooms.

The (Post)Modern World

Compulsory education is the modern world's answer to the problem of enculturation. We, collectively, annex the memetic engineering of our offspring by appointing a professional class of Idea-mongers to do the work for us. But it wasn't always this way. For 90 percent of our species' time on Earth, we were hunter-gatherers living in tribes. The tribal solution to education is for children to spend as much time as possible with the oldest members of the village. In an unchanging world, theirs would be the minds most worth copying. They know all the special skills and all the instructive stories. Their great longevity even enables them to work as conduits for vital information about natural disasters they may never have witnessed themselves. The 2004 tsunami delivered a thirty-foot-high wave to the Andaman Islands in the Indian Ocean. Indian government officials feared that the isolated tribes on the islands would suffer heavy casualties, but this wasn't the case: the tribes' oral traditions told them to escape to high ground if the sea retreated. While recent settlers suffered terribly, the tribespeople found safety in the hills.

The value of old age in our traditional lifestyle is why, many contend, our species has such a desperately long period of postreproductive life. No other animal or plant enjoys the grandparent years as much as we do, because in normal circumstances, infertile and infirm individuals are a bad idea: they are at least a burden and at worst unnecessary rivals

---

*In 1824 a selection of Ideas etched in stone (the Rosetta Stone, to be precise) achieved an incredible resurrection, after a dormancy of 2,018 years, in the mind of Frenchman Jean-François Champollion. They had formerly been alive in the minds of the priests of the Egyptian king Ptolemy V in 196 BC, and then endured, mindlessly, for all those years until the Frenchman, whose education had gifted him with a knowledge of how to read ancient Greek, added the water. The Ideas were in fact instructions on how to erect statues in temples, and not very useful per se to Champollion, but that doesn't matter; the cryonic preservation had worked. Through Champollion, those ancient memes were broadcast to the modern world. (As it happened, Champollion did something more interesting, memetically, with the stone. He used it to translate Egyptian hieroglyphs for the first time, uncorking a whole realm of other ancient memes.)

in a competitive world. But in our species they are vital because they instigate what cultural studies calls vertical transmission, the passage of Ideas "down" through the generations. In tribal times, this was the main process of information transfer at work in any human population, and indeed, until only very recently, vertical transmission ruled; compulsory education and the general social norms made sure of that. But the modern world appears to be characterized by a change in transfer practice. Today's youth learn from one another, via "horizontal transmission," as never before. Young people no longer need the elderly to survive. The wisdom of old age has lost much of its value because we no longer live in an unchanging world. The rate of change in Ideas, and the rate at which new Ideas arrive, goes up and up and up. Nowadays, even five-year-old Ideas are considered radically out-of-date. Some would say that this was due to a new capitalist agenda working to keep up our consumption of clothes, electronics, and the media, but I think that the rise of technology, as an after-effect of the rise of science, is the chief causal factor. Yes, one can imagine fast Ideas evolving to exploit the marketplace, but it is only as a result of revolutions in the transmission of Ideas that they have the opportunity to do so. The printing press, Bell's telephone, the radio, the television, the computer, the Internet, and the cell phone have all ratcheted meme Life up through design space. Now any Idea can be handed over to anyone at any time. It is open season for meme life. Cultural evolution has been thrown into overdrive, radiating left, right, and center as a multitude of Idea species that are designed to live fast, die young, and exploit the modern modus operandi of their chatting, hyperlinking, tweeting hosts.

Examples of these new *Homo sapiens* file into the café now as the Lido begins to fill up midmorning. Their look, attitude, lifestyle, and conversation are all dripping with the "new." Not that Ideas from the past are taboo—this dated café, with its Elvis pictures and soda fountain, is proof enough of that—just as long as those participating in the leading edge of culture don't consider it to be "old."

As they each approach the counter they make a choice of what brand of drink they will consume. Most say "Coke." That's because Coke tastes good and is refreshing, but it's also because Coke is "cool." Experiments with MRI scanners have shown that even though the reward centers of the brain are most stimulated by Pepsi, and in blind tastings, Pepsi is preferred on average, people will still say they prefer Coke if they can see the label.[2] Why? When subjects are placed under an MRI at the point that they see the label, an area just behind their forehead, the medial prefrontal cortex, starts launching fireworks. This area is self central. It is the part that fires when we try to relate to ourselves or establish our own identity. So it appears that the reason we like Coke more even though it tastes worse is that our self associates with the brand in some way. Coke, by accident or marketing purpose, has become part of our self-construction, our personality.

This is a triumph of meme Life over gene Life. Our genes have built us to prefer Pepsi, but we override this preference *without our own knowledge* as a result of the symbolic associations that the name and color of a can or bottle of Coke have inside our neocortex. Similar power struggles occur throughout the consumer world. When a study group was shown a range of branded items and celebrities that had been preselected by a gang of trendy design students to represent the spectrum of "cool" to "uncool," the medial prefrontal cortex was again busy, busy, together with a region called Brodmann's area 10. When analyzing these reactions, researchers at Caltech[3] noticed that the subjects fell into three response groups. The first had few strong reactions throughout. The second had a lot of activity in the medial prefrontal cortex when shown the cool brands and celebrities.* The third reacted strongly in Brodmann's area 10 to the uncool. Their conclusion was that some people don't take things too seriously, some people identify themselves with cultural artifacts that are communally regarded as cool, while others are terrified of the uncool.

Perhaps it is members of the second group who forge ahead on all of our accounts, categorizing which new Ideas are hot and which are not, and then it is the third group's paranoia to avoid the "not" that secures the place of a growing epidemic trend-Idea in the noosphere. Certainly there have been any number of Idea-based crazes during my lifetime, and I suspect that we, and now our societies, are built in such a way that there always will be. But isn't all this fashion a waste of time, materials, and money? Shouldn't we all be happy with being neither cool nor uncool, and spend our time, materials, and money on more useful pursuits, such as achieving world peace?

---

*I think it's fascinating that celebrities have a "product status" in our brains. Recent research suggests that in addition to neurons that flash the perception of brands such as Starbucks, we have others that work only when we perceive Jennifer Aniston.

Well, yeah, but when has that ever stopped us? We are meme monkeys. What makes us *happy* is Ideas—the memes have made sure of it—and they've done so by commandeering our reward centers. The very best strategy for any Idea to guarantee long life and high fecundity is to give us, their hosts, a sprinkle of reward anytime we entertain them. Our reward systems were originally set up by our genes so we'd be encouraged to perform acts that benefitted *them*: engaging in sexual intercourse, drinking water when dehydrated, taking care of our offspring. But the memes have clearly hijacked the system. How else would we derive pleasure from passing on Ideas—teaching our friends a new golf grip, telling a joke, adding a factoid to a dinner conversation, starting any sentence with "in my opinion"? That group of people in the Caltech test unconcerned with the coolness of products—they may not care much about fashion labels, but I'll bet they do get pleasure from using other types of Ideas. Something must float their boats: maybe stamp collecting or World of Warcraft or baseball. We all have our vices, and that explains how such a variety of Ideas can find niches in the cultural jungle we've built. We don't need to be embarrassed by this status quo. We should celebrate it. We love Ideas. So what?

The Truth

But this doesn't necessarily mean that we should encourage all Ideas. Generally in Life, diversity is good, and memetic diversity is good for meme Life—it can only assist in its aim to explore all of design space— but that doesn't mean to say that it's good for *us*. History tells us that we need to be more discerning with our memes.

At the heart of the issue is the fact that the process of succession is not preordained. It is well understood in ecology that both the external environment and the composition of the species will readily "deflect" succession to a different outcome. So it is with the Idea succession that takes place in the growing minds of our children. Children are vulnerable. They will gaily inherit any truth. The cultural environment inside and outside schools has a profound impact on the direction of succession in a child's mind, as does the nature of the species that join the community. I had a typical Western education. Rationalism and liberalism were its chief flavors, and hence my mind community tastes of them, too. But not everyone has an education like mine. Thanks to our fallibilities in choosing exactly what Ideas to believe, we all inevitably become hosts to untruths, biases, and bad Ideas. We are all vulnerable to ideologies, or perhaps we should call them Idea ecologies. Governments (including mine) both consciously and subconsciously foster Idea communities in their children that suit their sociopolitical aims. We are all subject to memetic engineering. And this alone can explain a lot of our species' historical failures. The difference in the behavior of a young German in 1940 and a young German in 1980 is purely down to differences in their cultural environments and the floating stock of Idea species they encountered. We are all born meme monkeys and are all vulnerable as such.

Now that these goggles are available, goggles that allow us to see clearly the water around us, my hope is that we can become aware of this frailty and compensate for it. Let me give you an example. Among the troubled mountains of the Caucasus lie two tiny nations, Ingushetia and North Ossetia, who, despite the fact that their names sound quite sweet together, have been bitter foes for as long as either side can remember. The Ingush think that the North Ossetians started it. The North Ossetians think it was the Ingush. If you go to an Ingush school, you'll be taught the inhuman violence of the North Ossetians. If you go to a North Ossetian school, you'll get a lesson in the treachery of the barbaric Ingush. It was ever thus, because the minds of Ingush and Ossetian children are easy prey for these self-destructive inherited Ideas.

Recently, however, a group of charity workers have had their own Idea. They've decided to write a joint Ingush-Ossetian history book. Instead of being slaves to memes, they would make memes work for them. They would do some memetic engineering. They would compare and contrast the two received wisdoms and, using historical evidence from outside the immediate area, compile a consensus history that, of course, would be far more tempered. Teachers in both countries, they guessed, would spit as they taught the new history, but teach it they must. It might take a decade for the primary successions to be completed, but eventually, with enough support, the next generation of Caucasians would be unshackled from the bad Ideas of their past and enter adulthood far more inclined to forgive, forget, and look up from their provincial lives, beyond, to a wider world where seven billion other humans wrestle with the burden of their own genes and memes, desperately trying to make some progress in a mindless world.

Lord knows there are hundreds if not thousands of similar situations in play. Gaza strips, credit crunches, suicide bombers, Zimbabwean dollars—all the product of underlying toxic Ideas. Gene therapy requires dust-free laboratories and harnessed retroviruses and cationic dendrimers. If meme therapy can be administered with nothing more complex than a new history book, surely there is a chance that we can use our new understanding of ourselves to attend to these manmade disasters.

Ads and I leave the Lido, sated, and walk up onto the bridge of the Bow River, overlooking a park on the other bank on the edge of downtown. It's a sunny Saturday morning and the meme monkeys of Calgary are out in force—flying kites, setting up picnics, boating, jogging, reading the paper, searching through the garbage bins—old ones, young ones, middle-aged ones, some with kids, some "dual-income, no-kids," but none of them weirdoes. I see that now; they are all simply performing their given role in Darwin's universe, the design goal of the most remarkable evolutionary project that Life has ever undertaken.

The river is green and fat below us, smacking the concrete abutments. After a while we spot trout in the reeds in the shadow of the arches. They conquer the flow without any apparent effort. They skip to the left and to the right as the water confronts them, perfectly adapted to combat its idiosyncrasies. They are aware of the water that surrounds them. Every ounce of their being is built to command it. They refuse to go with the flow. Instead, they swim.

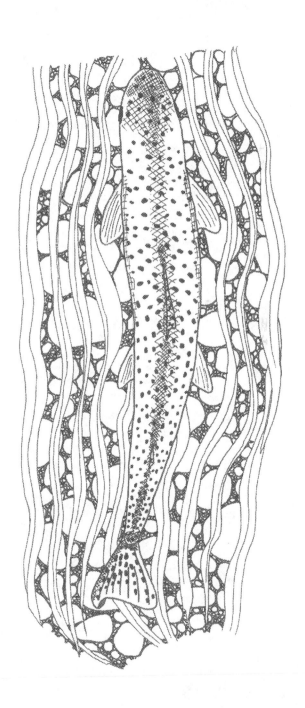

Notes

Chapter 3: Evolution, Minnesota

1. Daniel Dennett, *Darwin's Dangerous Idea*, 1995, New York : Simon & Schuster, p. 50.

Chapter 6: Selection, Wyoming

1. Richard Dawkins, *The Greatest Show on Earth*, 2009, New York: Free Press, p. 73.

Chapter 9: How the West Was Won II

1. Quoted in Jacques Hadamard's *The Psychology of Inventing in the Mathematical Field*, Princeton: Princeton University Press, 1949, p. 16.

Chapter 11: A Beginner's Guide to Tepee Taxonomy

1. S. K., Huber, L. F. De León, A. P. Hendry, E. Bermingham, and J. Podos, "Reproductive Isolation of Sympatric Morphs in a Population of Darwin's Finches," *Proc Biol Sci* 274, no. 1619 (July 22, 2007): 1709–14.
2. John McWhorter, *The Power of Babel*, London: Arrow, 2001, p. 13.
3. www.ethnologue.com (October 2010).

Chapter 12: Bound by Imagination

1. See Daniel Gilbert's *Stumbling on Happiness* (2007).

Chapter 14: The Past

1. Roy Mukamel et al., "Single-Neuron Responses in Humans during Execution and Observation of Actions," *Current Biology* 20, no. 8 (2010).
2. Xu J., P. J. Gannon, K. Emmorey, J. F. Smith, and A. R. Braun, "Symbolic Gestures and Spoken Language Are Processed by a Common Neural System," *Proc Natl Acad Sci USA* 106 (2009): 20664–69.
3. Chun Siong Soon, Marcel Brass, Hans-Jochen Heinze, and John-Dylan Haynes, "Unconscious Determinants of Free Decisions in the Human Brain," *Nature Neuroscience* 11 (2008): 543–45.
4. E. B. Renaud, *The Archaeological Survey of the High Western Plains*, Denver, University of Denver, 1936.

Chapter 15: The Present

1. Quoted in T. C. Whitmore, *An Introduction to Tropical Rain Forests*, Oxford: Clarendon Press, 1990.
2. McClure et al. "Neural Correlates of Behavioral Preference for Culturally Familiar Drinks," *Neuron* 44 (2004): 379–87.
3. See the Quartz research group, www.qlab.caltech.edu / Qpeople.htm.

Bibliography

The Goggle Makers

I've made a conscious effort to avoid naming scientists and philosophers throughout the text unless it's really important to do so. My justification is that it formalizes and slows down the story. However, *now* it really *is* important to do so. So, who made my new-world-view goggles?

The first to scrabble together the raw materials were those that first saw parallels between biological and cultural evolution. Darwin himself was key among these. He was always of the opinion that natural selection should be considered "substrate-neutral," applicable to theaters of operation beyond just organisms, and he used the competition between words and tool designs and the evolution of languages to illustrate his theory of natural selection. His contemporary Herbert Spencer was even more committed to aligning the two evolutions, and regarded the culture of civilizations as a fundamental evolutionary environment in which all human minds grow. William James, the pioneering psychologist, drew the same conclusion and, in 1880, called for his followers to gain a thorough understanding of Darwinism such that they could apply it to explain the processes in human societies. But, these early theorists were constrained in one significant respect: none of them knew of the gene, so they could hardly take the meme's-eye view.

Darwin, C. (1859/1964). *The Origin of Species*. 1st ed. With an Introduction by Ernst Mayr. Cambridge, Mass.: Harvard University Press.

———. (1877/2004). *The Descent of Man*. 2nd ed. With an introduction by Adrian Desmond and James Moore. London: Penguin.

James, W. (1880). "Great Men, Great Thoughts, and the Environment." *Atlantic Monthly* 66: 441–59.

Spencer, H. (1855). *The Principles of Psychology*. London: Longman, Brown, Green and Longmans.

The road to the meme proper begins with a German zoologist called Richard Wolfgang Semon, who was the first in modern times to posit that culture could be present as millions of units in the collective human memory. In a paper in 1904, he called these units of "memory-feeling" mnemes, after the Ancient Greek muse of memory, and suggested that when the brain perceives a stimulus, a psychological state composed of an "alteration of nerves" is formed in the memory. This state, he proposed, could be revived should the same stimulus appear again.

Semon, R. W. (1921). *The Mneme*. London: George Allen and Unwin.

After Semon, things went very quiet until the dawn of the gene's-eye view. It was Dawkins who, in publicizing Hamilton and Williams's take on Life in his breakthrough book, *The Selfish Gene*, rekindled the discussion with a final chapter called "Memes: The New Replicators." In common with Darwin, Dawkins wanted to assert that natural selection was substrate-neutral. To do so, he chose to propose a cultural replicator, the "meme," coming up with the word without prior knowledge of Semon's text, merely because it denoted something to do with the memory and rhymed-*ish* with the word *gene*. It was a good choice. As the inventor of meme theory, Dawkins was well aware that a word able to prefix others such as with *memetics*, *memone*, *memeplex* and *memotype*, was bound to stick.

Dawkins, R. (1976). *The Selfish Gene*. Oxford: Oxford University Press.

Dawkins's meme theory was bold and brash, and not to everyone's taste. In the 1980s a series of researchers tried to get the same smug explanatory power by lashing our cultural evolution ever so tightly to our biological evolution in complex "coevolutionary models." The sociobiologist E. O. Wilson, the leading human geneticist Luigi Luca Cavalli-Sforza, and Californian anthropologists Robert Boyd and Peter Richerson all published work that admitted that cultural evolution tasted Darwinian but that wouldn't commit to a cultural replicator.

Boyd, R., and P. Richerson. (1985). *Culture and the Evolutionary Process*. Chicago: University of Chicago Press.

Cavalli-Sforza, L., and M. Feldman. (1981). *Cultural Transmission and Evolution: A Quantitative Approach*. Princeton, N.J.: Princeton University Press.

Lumsden, C., and E. O. Wilson. (1981). *Genes, Mind and Culture: The Coevolutionary Process*. Cambridge, Mass.: Harvard University Press.

In the 1990s the discussion was temporarily interrupted by the glitzy arrival of a brand-new science: evolutionary psychology. This project attempted to explain human behaviors by reducing them back to the beneficial genes that coded for them. Evolutionary psychology was greeted with great excitement, and the fervor of research in this discipline was one of the most notable activities in science in the 1990s, but ultimately it failed to deliver on its promises. The reductionism of some evolutionary psychologists was too greedy, their genetic determinism too profound, and it began to become associated with blinkered science. For example:

Pinker, S. (1994). *The Language Instinct*. New York: W. Morrow and Co.

———. (1997). *How the Mind Works*. New York: W.W. Norton.

Tooby, J., and L. Cosmides. (1992). "The Psychological Foundations of Culture." In J. Barkow, L. Cosmides, and J. Tooby (eds.), *The Adapted Mind: Evolutionary Psychology and the Generation of Culture*. Oxford: Oxford University Press, pp. 19–136.

It was in the middle of all this that American philosopher of science Daniel Dennett wrote *Darwin's Dangerous Idea*. In it he claimed that, from a philosophical point of view, we shouldn't be satisfied with any description of our evolution that is *unDarwinian*: Darwinism is the only thing we know of that brings design mindlessly, so we shouldn't rest until we understand the Darwinism that led to us. He suggested going back to Dawkins's memes and giving them another chance. At first he couldn't image how we could build such a science, but his work since has implored us to try harder.

Dennett, D. C. (1995). *Darwin's Dangerous Idea: Evolution and the Meanings of Life.* New York: Simon & Schuster.

And many have taken up the challenge. Notable researchers working today in this area include psychologists Susan Blackmore, Alex Mesoudi, and Kevin Laland; anthropologist Robert Aunger; and social biologist Andrew Whiten. As the sciences that study the structure and function of our brains, neurology, and neuropsychology mature, this, I suspect, will be just the beginning.

Aunger, R. (2002). *The Electric Meme: A New Theory of How We Think.* New York: Free Press.

Blackmore, S. (1999). *The Meme Machine.* Oxford: Oxford University Press.

Mesoudi, A., A. Whiten, and K. N. Laland. (2006). "Towards a Unified Science of Cultural Evolution." *Behavioral and Brain Sciences* 29(4): 329–83.

Additional Reading

There follows a list of other texts that relate closely to the content of this book. All have played their part in informing the worldview through my particular set of goggles.

Aunger, R. (2000). *Darwinizing Culture: The Status of Memetics as a Science.* Oxford: Oxford University Press.

Begon, M., C. R. Townsend, and J. L. Harper. (2006). *Ecology: From Individuals to Ecosystems.* 4th ed. Malden, Mass.: Blackwell Pub.

Brodie, R. (1996). *Virus of the mind: The New Science of the Meme*. Seattle, Wash.: Integral Press.

Cavalli-Sforza, L. L. (2000). *Genes, Peoples, and Languages*. London: Penguin Books.

Darwin, C. (1959). *The Voyage of the Beagle*. New York: Harper.

Dawkins, R. (1986). *The Blind Watchmaker*. New York: W.W. Norton.

———. (2004). *The Ancestor's Tale: A Pilgrimage to the Dawn of Evolution*. Boston: Houghton Mifflin.

———. (2009). *The Greatest Show on Earth: The Evidence for Evolution*. New York: Free Press.

Dennett, D. C. (2003). *Freedom Evolves*. New York: Viking.

———. (2006). *Breaking the Spell: Religion as a Natural Phenomenon*. New York: Viking.

Diamond, J. M. (1998). *Guns, Germs, and Steel: The Fates of Human Societies*. New York: W.W. Norton and Co.

———. (2005). *Collapse: How Societies Choose to Fail or Succeed*. New York: Viking.

Distin, K. (2005). *The Selfish Meme: A Critical Reassessment*. Cambridge, UK: Cambridge University Press.

Dunbar, R.I.M., C. Knight, and C. Power. (1999). *The Evolution of Culture: An Interdisciplinary View*. New Brunswick, N.J.: Rutgers University Press.

Evans, D., R. Appignanesi, and O. Zarate. (2005). *Introducing Evolutionary Psychology*. Cambridge, Mass.: Icon Books.

Fox, K. (2004). *Watching the English: The Hidden Rules of English Behavior*. London: Hodder and Stoughton.

Frank, L. (2009). *Mindfield: How Brain Science Is Changing Our World*. Oxford: Oneworld.

Gilbert, D. T. (2006). *Stumbling on Happiness*. New York: Knopf.

Gladwell, M. (2000). *The Tipping Point: How Little Things Can Make a Big Difference*. Boston: Little, Brown.

Goble, P. (2007). *Tipi: Home of the Nomadic Buffalo Hunters*. Bloomington, Ind.: World Wisdom.

Grafen, A., and M. Ridley. (2006). *Richard Dawkins: How a Scientist Changed the Way We Think: Reflections by Scientists, Writers, and Philosophers*. Oxford: Oxford University Press.

Iacoboni, M. (2008). *Mirroring People: The New Science of How We Connect with Others*. New York: Farrar, Straus and Giroux.

Jablonka, E. and M. J. Lamb. (2005). *Evolution in Four Dimensions: Genetic, Epigenetic, Behavioral, and Symbolic Variation in the History of Life*. Cambridge, Mass.: MIT Press.

Jones, S. (2000) *Almost Like A Whale: "The Origin of Species" Updated*. London: Anchor.

Laland, K. N. and B. G. Galef. (2009). *The Question of Animal Culture*. Cambridge, Mass.: Harvard University Press.

Laubin, R., G. Laubin, and S. Vestal. (1997). *The Indian Tipi: Its History, Construction, and Use*. Norman: University of Oklahoma Press.

Lynch, A. (1996). *Thought Contagion: How Belief Spreads Through Society*. New York: Basic Books.

Lynch, G. and R. Granger. (2008). *Big Brain: the Origins and Future of Human Intelligence*. New York: Palgrave Macmillan.

McWhorter, J. H. (2003). *The Power of Babel: A Natural History of Language*. London: Arrow.

Marcus, G. F. (2008). *Kluge: The Haphazard Construction of the Human Mind*. Boston: Houghton Mifflin.

Mesoudi, A. (2005). *The Transmission and Evolution of Human Culture*. St. Andrews: University of St. Andrews.

Raby, P. (2001). *Alfred Russel Wallace: A Life*. Princeton, N.J.: Princeton University Press.

Richerson, P. J. and R. Boyd. (2005). *Not by Genes Alone: How Culture Transformed Human Evolution*. Chicago: University of Chicago Press.

Ridley, M. (2003). *Nature via Nurture: Genes, Experience, and What Makes Us Human*. New York: HarperCollins.

Schilthuizen, M. (2001). *Frogs, Flies and Dandelions: Speciation—The Evolution of New Species*. Oxford: Oxford University Press.

Sigmund, K. (1993). *Games of Life: Explorations in Ecology, Evolution, and Behaviour*. Oxford: Oxford University Press.

Smith, J., and E. Szathmáry. (1999). *The Origins of Life: From the Birth of Life to the Origin of Language*. Oxford: Oxford University Press.

Waldman, C. and M. Braun. (1988). *Encyclopedia of Native American Tribes*. New York: Facts on File.

Watson, P. (2005). Ideas: *A History of Thought and Invention, from Fire to Freud*. New York: HarperCollins.

Wilson, E. O. (1998). *Consilience: The Unity of Knowledge*. New York: Knopf.

Index

About the Author

JONNIE HUGHES spent his childhood methodically scouring the rock pools of the Devon coast in South West England. Captivated by the miniature wilderness he discovered, he studied ecology and evolution at the University of Leeds, then moved to London to build a career telling others all about Life, what it is and how it works. He taught it in college, wrote about it in newspapers and magazines, and made films about it for the BBC, Discovery, and National Geographic Channel. This is his first book.

Printed in the United States
By Bookmasters